REA: THE LEADER IN CLEP® TEST PREP

CLEP® NATURAL SCIENCES

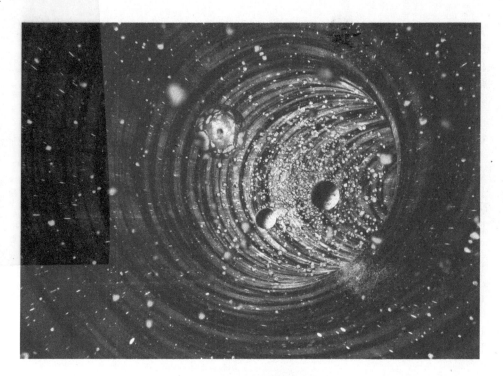

Laurie Ann Callihan
University Fellow
Florida State University
Tallahassee, Florida

David Callihan
Physical Science and Math Teacher
Bishop Hall Charter School
Thomasville, Georgia

 Research & Education Association
Visit our website at: www.rea.com

Research & Education Association
61 Ethel Road West
Piscataway, New Jersey 08854
E-mail: info@rea.com

CLEP® Natural Sciences with Online Practice Exams

Copyright © 2016 by Research & Education Association, Inc.
Prior editions copyright © 2013, 2010 by Research & Education Association, Inc. All rights reserved. No part of this book may be reproduced in any form without permission of the publisher.

Printed in the United States of America

Library of Congress Control Number 2016934662

ISBN-13: 978-0-7386-1207-2
ISBN-10: 0-7386-1207-3

LIMIT OF LIABILITY/DISCLAIMER OF WARRANTY: Publication of this work is for the purpose of test preparation and related use and subjects as set forth herein. While every effort has been made to achieve a work of high quality, neither Research & Education Association, Inc., nor the authors and other contributors of this work guarantee the accuracy or completeness of or assume any liability in connection with the information and opinions contained herein and in REA's software and/or online materials. REA and the authors and other contributors shall in no event be liable for any personal injury, property or other damages of any nature whatsoever, whether special, indirect, consequential or compensatory, directly or indirectly resulting from the publication, use or reliance upon this work.

CLEP® is a registered trademark of the College Board, which was not involved in the production of, and does not endorse, this product. All other trademarks cited in this publication are the property of their respective owners.

Cover image: © istockphoto.com/gremlin

 REA® is a registered trademark of
Research & Education Association, Inc.

CONTENTS

About Our Authors.. vi
About REA.. vii

CHAPTER 1
Passing the CLEP Natural Sciences Exam 1
    Getting Started.. 3
    The REA Study Center.. 4
    An Overview of the Exam 5
    All About the CLEP Program.. 6
    Options for Military Personnel and Veterans........................ 8
    SSD Accommodations for Candidates with Disabilities 8
    6-Week Study Plan 9
    Test-Taking Tips 9
    The Day of the Exam.. 10
Online Diagnostic Test .... *www.rea.com/studycenter*

PART I—BIOLOGICAL SCIENCE

CHAPTER 2
Evolution and Classification 13
    The Origin of Life 16
    Evolution of Life 19
    Mechanisms of Evolution... 23
    Mechanisms of Speciation... 26
    Classification of Living Organisms 28

CHAPTER 3
Cellular and Molecular Biology... 33
    The Structure and Function of Cells................................... 35
    Chemical Nature of the Gene 48

CHAPTER 4
Structure and Function of Plants and Animals; Genetics 61
- Plants (Botany) 63
- Animals (Zoology) 69
- Principles of Heredity (Genetics) 83

CHAPTER 5
Ecology and Population Biology 91
- Ecology 93
- Population Growth and Regulation.. 99

PART II—PHYSICAL SCIENCE

CHAPTER 6
Atomic Chemistry 107
- Structure of the Atom 109
- Nuclear Reactions and Equations... 112
- Rate of Decay; Half-Life 114

CHAPTER 7
Chemistry of Reactions... 115
- Common Elements 117
- Chemical Bonds.. 118
- Chemical Reactions 119

CHAPTER 8
Physics 121
- Heat.. 123
- The Laws of Thermodynamics 124
- States of Matter... 124
- Classical Mechanics 129
- Theory of Relativity 132

CHAPTER 9
Energy **135**
    Electricity and Magnetism... 137
    Waves: Sound and Light 139

CHAPTER 10
The Universe **143**
    Astronomy 145

CHAPTER 11
Earth **155**
    Atmosphere... 157
    Earth's Layers 160

Practice Test 1 (also available online at *www.rea.com/studycenter*)............... 165
    Answer Key 193
    Detailed Explanations of Answers 194

Practice Test 2 (also available online at *www.rea.com/studycenter*) 209
    Answer Key 239
    Detailed Explanations of Answers 240

Answer Sheets... 257

Glossary 261

Index 271

ABOUT OUR AUTHORS

Laurie Ann Callihan, Ph.D., received her Bachelor of Science in Biology from San Diego Christian College in El Cajon, California. She received her Ph.D. in Science Education as a University Fellow at Florida State University. She has extensive writing, speaking, and teaching experience. Her publications include REA's *CLEP Biology with Online Exams* and other REA titles and journal articles. She is an internationally known speaker on education, having been the keynote speaker at state and local education conventions throughout the U.S., Canada, and Mexico. In addition to years of study, curriculum development, and teaching Biology and Physical Science, her current research focuses on development of effective teaching methods and tools for conceptual understanding.

David Callihan earned a Bachelor of Science in Geophysics from San Diego Christian College in El Cajon, California and a Master's in Science Education at Florida State University. He currently teaches physical science and mathematics at Bishop Hall Charter School in Thomasville (Thomas County), Georgia. He began his teaching career at DeSoto Middle School in Arcadia, Florida, where he was recognized as the top-performing teacher based on AYP results. Prior to becoming a teacher, Mr. Callihan had a successful career in manufacturing, oil exploration, and high technology sales. Mr. Callihan is a Zero-G teacher coach, and has worked with the Florida Space Grant Consortium, the Center for Oceanic Studies in Educational Excellence, and the Florida Institute of Phosphate Research.

ABOUT REA

Founded in 1959, Research & Education Association (REA) is dedicated to publishing the finest and most effective educational materials—including study guides and test preps—for students of all ages.

Today, REA's wide-ranging catalog is a leading resource for students, teachers, and other professionals. Visit *www.rea.com* to see a complete listing of all our titles.

ACKNOWLEDGMENTS

We would like to thank Pam Weston, Publisher, for setting the quality standards for production integrity and managing the publication to completion; John Paul Cording, Vice President, Technology, for coordinating the design and development of the REA Study Center; Larry B. Kling, Director of Editorial Services, for his assistance in bringing this book to market; Diane Goldschmidt, Managing Editor, for coordinating development of this edition; Kathy Caratozzolo of Caragraphics for typesetting; Ellen Gong for proofreading; and Terry Casey for indexing.

CHAPTER 1

Passing the CLEP Natural Sciences Exam

CHAPTER 1

PASSING THE CLEP NATURAL SCIENCES EXAM

Congratulations! You're joining the millions of people who have discovered the value and educational advantage offered by the College Board's College-Level Examination Program, or CLEP. This test prep focuses on what you need to know to succeed on the CLEP Natural Sciences exam, and will help you earn the college credit you deserve while reducing your tuition costs.

GETTING STARTED

There are many different ways to prepare for a CLEP exam. What's best for you depends on how much time you have to study and how comfortable you are with the subject matter. To score your highest, you need a system that can be customized to fit you: your schedule, your learning style, and your current level of knowledge.

This book, and the online tools that come with it, allow you to create a personalized study plan through three simple steps: assessment of your knowledge, targeted review of exam content, and reinforcement in the areas where you need the most help.

Let's get started and see how this system works.

Test Yourself and Get Feedback	Assess your strengths and weaknesses. The score report from your online diagnostic exam gives you a fast way to pinpoint what you already know and where you need to spend more time studying.
Review with the Book	Armed with your diagnostic score report, review the parts of the book where you're weak and study the answer explanations for the test questions you answered incorrectly.
Ensure You're Ready for Test Day	After you've finished reviewing with the book, take our full-length practice tests. Review your score reports and re-study any topics you missed. We give you two full-length practice tests to ensure you're confident and ready for test day.

THE REA STUDY CENTER

The best way to personalize your study plan is to get feedback on what you know and what you don't know. At the online REA Study Center, you can access two types of assessment: a diagnostic exam and full-length practice exams. Each of these tools provides true-to-format questions and delivers a detailed score report that follows the topics set by the College Board.

Diagnostic Exam

Before you begin your review with the book, take the online diagnostic exam. Use your score report to help evaluate your overall understanding of the subject, so you can focus your study on the topics where you need the most review.

Full-Length Practice Exams

Our full-length practice tests give you the most complete picture of your strengths and weaknesses. After you've finished reviewing with the book, test what you've learned by taking the first of the two online practice exams. Review your score report, then go back and study any topics you missed. Take the second practice test to ensure you have mastered the material and are ready for test day.

If you're studying and don't have Internet access, you can take the printed tests in the book. These are the same practice tests offered at the REA Study Center, but without the added benefits of timed testing conditions and diagnostic score reports. Because the actual exam is Internet-based, we recommend you take at least one practice test online to simulate test-day conditions.

AN OVERVIEW OF THE EXAM

The CLEP Natural Sciences exam consists of approximately 120 multiple-choice questions, each with five possible answer choices, to be answered in 90 minutes.

The exam covers the material one would find in a freshman or sophomore general science survey course covering biology and physical science. It is meant for students who are non-science majors. The exam stresses basic facts and principles, as well as general theoretical approaches used by scientists.

The approximate breakdown of topics is as follows:

50% Biological Science

- 10% Origin and evolution of life, classification of organisms
- 10% Cell organization, cell division, chemical nature of the gene, bioenergetics, biosynthesis
- 20% Structure, function, and development in organisms; patterns of heredity
- 10% Concepts of population biology with emphasis on ecology

50% Physical Science

- 7% Atomic and nuclear structure and properties, elementary particles, nuclear reactions
- 10% Chemical elements, compounds and reactions, molecular structure and bonding
- 12% Heat, thermodynamics, and states of matter; classical mechanics; relativity
- 4% Electricity and magnetism, waves, light, and sound
- 7% The universe: galaxies, stars, the solar system
- 10% The Earth: atmosphere, hydrosphere, structure features, geologic processes, and history

ALL ABOUT THE CLEP PROGRAM

What is the CLEP?

CLEP is the most widely accepted credit-by-examination program in North America. The CLEP program's 33 exams span five subject areas. The exams assess the material commonly required in an introductory-level college course. Based on recommendations from the American Council on Education, examinees can earn from three to nine credits at more than 2,900 colleges and universities in the U.S. and Canada. For a complete list of the CLEP subject examinations offered, visit the College Board website: *www.collegeboard.org/clep*.

Who takes CLEP exams?

CLEP exams are typically taken by people who have acquired knowledge outside the classroom and who wish to bypass certain college courses and earn college credit. The CLEP program is designed to reward examinees for learning—no matter where or how that knowledge was acquired.

Although most CLEP examinees are adults returning to college, many graduating high school seniors, enrolled college students, military personnel, veterans, and international students take CLEP exams to earn college credit or to demonstrate their ability to perform at the college level. There are no prerequisites, such as age or educational status, for taking CLEP examinations. However, because policies on granting credits vary among colleges, you should contact the particular institution from which you wish to receive CLEP credit.

How is my CLEP score determined?

Your CLEP score is based on two calculations. First, your CLEP raw score is figured; this is just the total number of test items you answer correctly. After the test is administered, your raw score is converted to a scaled score through a process called *equating*. Equating adjusts for minor variations in difficulty across test forms and among test items, and ensures that your score accurately represents your performance on the exam regardless of when or where you take it, or on how well others perform on the same test form.

Your scaled score is the number your college will use to determine if you've performed well enough to earn college credit. Scaled scores for the CLEP exams are delivered on a 20-80 scale. Institutions can set their own scores for granting college credit, but a good passing estimate (based on recommendations from the American Council on Education) is generally a scaled score of 50, which usually requires getting roughly 66% of the questions correct.

For more information on scoring, contact the institution where you wish to be awarded the credit.

Who administers the exam?

CLEP exams are developed by the College Board, administered by Educational Testing Service (ETS), and involve the assistance of educators from throughout the United States. The test development process is designed and implemented to ensure that the content and difficulty level of the test are appropriate.

When and where is the exam given?

CLEP exams are administered year-round at more than 1,200 test centers in the United States and can be arranged for candidates abroad on request. To find the test center nearest you and to register for the exam, contact the CLEP Program:

CLEP Services
P.O. Box 6600
Princeton, NJ 08541-6600
Phone: (800) 257-9558 (8 a.m. to 6 p.m. ET)
Fax: (610) 628-3726
Website: *www.collegeboard.org/clep*

The CLEP iBT Platform

To improve the testing experience for both institutions and test-takers, the College Board's CLEP Program has transitioned its 33 exams from the eCBT platform to an Internet-based testing (iBT) platform. All CLEP test-takers may now register for exams and manage their personal account information through the "My Account" feature on the CLEP website. This new feature simplifies the registration process and automatically downloads all pertinent information about the test session, making for a more streamlined check-in.

OPTIONS FOR MILITARY PERSONNEL AND VETERANS

CLEP exams are available free of charge to eligible military personnel and eligible civilian employees. All the CLEP exams are available at test centers on college campuses and military bases. Contact your Educational Services Officer or Navy College Education Specialist for more information. Visit the DANTES or College Board websites for details about CLEP opportunities for military personnel.

Eligible U.S. veterans can claim reimbursement for CLEP exams and administration fees pursuant to provisions of the Veterans Benefits Improvement Act of 2004. For details on eligibility and submitting a claim for reimbursement, visit the U.S. Department of Veterans Affairs website at *www.gibill.va.gov.*

CLEP can be used in conjunction with the Post-9/11 GI Bill, which applies to veterans returning from the Iraq and Afghanistan theaters of operation. Because the GI Bill provides tuition for up to 36 months, earning college credits with CLEP exams expedites academic progress and degree completion within the funded timeframe.

SSD ACCOMMODATIONS FOR CANDIDATES WITH DISABILITIES

Many test candidates qualify for extra time to take the CLEP exams, but you must make these arrangements in advance. For information, contact:

College Board Services for Students with Disabilities
P.O. Box 8060
Mt. Vernon, Illinois 62864-0060
Phone: (609) 771-7137 (Monday through Friday, 8 A.M. to 6 P.M. ET)
TTY: (609) 882-4118
Fax: (866) 360-0114
E-mail: ssd@info.collegeboard.org

6-WEEK STUDY PLAN

Although our study plan is designed to be used in the six weeks before your exam, it can be condensed to three weeks by combining each two-week period into one.

Be sure to set aside enough time—at least two hours each day—to study. The more time you spend studying, the more prepared and relaxed you will feel on the day of the exam.

Week	Activity
1	Take the Diagnostic Exam. The score report will identify topics where you need the most review.
2–4	Study the review chapters. Use your diagnostic score report to focus your study.
5	Take Practice Test 1 at the REA Study Center. Review your score report and re-study any topics you missed.
6	Take Practice Test 2 at the REA Study Center to see how much your score has improved. If you still got a few questions wrong, go back to the review and study any topics you may have missed.

TEST-TAKING TIPS

Know the format of the test. Familiarize yourself with the CLEP computer screen beforehand by logging on to the College Board website. Waiting until test day to see what it looks like in the pretest tutorial risks injecting needless anxiety into your testing experience. Also, familiarizing yourself with the directions and format of the exam will save you valuable time on the day of the actual test.

Read all the questions—completely. Make sure you understand each question before looking for the right answer. Reread the question if it doesn't make sense.

Read all of the answers to a question. Just because you think you found the correct response right away, do not assume that it's the best answer. The last answer choice might be the correct answer.

Work quickly and steadily. You will have 90 minutes to answer 120 questions, so work quickly and steadily. Taking the timed practice tests online will help you learn how to budget your time.

Use the process of elimination. Stumped by a question? Don't make a random guess. Eliminate as many of the answer choices as possible. By eliminating just two answer choices, you give yourself a better chance of getting the item correct, since there will only be three choices left from which to make your guess. Remember, your score is based only on the number of questions you answer correctly.

Don't waste time! Don't spend too much time on any one question. Your time is limited so pacing yourself is very important. Work on the easier questions first. Skip the difficult questions and go back to them if you have the time.

Look for clues to answers in other questions. If you skip a question you don't know the answer to, you might find a clue to the answer elsewhere on the test.

Be sure that your answer registers before you go to the next item. Look at the screen to see that your mouse-click causes the pointer to darken the proper oval. If your answer doesn't register, you won't get credit for that question.

THE DAY OF THE EXAM

On test day, you should wake up early (after a good night's rest, of course) and have breakfast. Dress comfortably, so you are not distracted by being too hot or too cold while taking the test. (Note that "hoodies" are not allowed.) Arrive at the test center early. This will allow you to collect your thoughts and relax before the test, and it will also spare you the anxiety that comes with being late.

Before you leave for the test center, make sure you have your admission form and another form of identification, which must contain a recent photograph, your name, and signature (i.e., driver's license, student identification card, or current alien registration card). You may wear a watch. However, you may not wear one that makes noise, because it may disturb the other test-takers. No cell phones, dictionaries, textbooks, notebooks, briefcases, or packages will be permitted, and drinking, smoking, and eating are prohibited.

Good luck on the CLEP Natural Sciences exam!

PART I
Biological Science

CHAPTER 2

Evolution and Classification

CHAPTER 2

EVOLUTION AND CLASSIFICATION

Darwinian Concept of Natural Selection

Current theories of evolution have their basis in the work of Charles Darwin and one of his contemporaries, Alfred Russell Wallace. Darwin's book, *The Origin of Species by Means of Natural Selection*, served to catalyze the study of evolution across scientific disciplines. A synopsis of Darwin's ideas follows.

From a study of populations, we know that population growth and maintenance of a species is dependent on limiting factors. Individuals within the species that are unable to acquire the minimum requirement of resources are unable to reproduce. The ecosystem can support only a limited number of organisms—known as the carrying capacity (K).

Once the carrying capacity is reached, a competition for resources ensues. Darwin considered this competition to be the basic *struggle for existence*. Some of the competitors will fail to survive. Within every population, there is variation among traits. Darwin proposed that those individuals who win the competition for resources pass those successful traits on to their children. Clearly, only the well adapted survive, and only the surviving competitors reproduce successfully and thus pass on traits generation after generation. Therefore, traits providing the competitive edge will be represented most often in succeeding generations.

Modern Concept of Natural Selection

Although the concepts of natural selection put forth by Darwin still form the basis of evolutionary theory today, Darwin had no real knowledge of genetics when he presented his ideas.

Several years after Darwin's writings, Mendel's work (on experimental genetics) was rediscovered independently by three scientists. The laws of genetics served to support the suppositions Darwin had made. Over the next 40 years, the study of genetics included not only individual organisms but also population genetics (how traits are preserved, changed, or introduced within a population of organisms). Progress in the studies of biogeography and paleontology of the early 1900s also served to reinforce Darwin's basic observations.

The modern concept of natural selection emerged from Darwin's original ideas, with additions and confirmations of genetics, population studies, and paleontology. This **modern synthesis** focused on the concept that evolution was a process of gradual adaptive change in traits among populations and communities (over thousands or hundreds of thousands of generations). Evolution was not a process of individual-by-individual change, but change that occurred in entire populations within communities due to major environmental events.

THE ORIGIN OF LIFE

Evolution of the First Cells

The modern theory of the evolution of life on Earth suggests that the earliest forms of life began approximately four billion years ago. Theories suggest that conditions on Earth were very different from conditions today. The atmosphere of the early Earth was likely chemically reducing (accepting electrons), composed mainly of methane (CH_4), ammonia (NH_3), hydrogen sulfide (H_2S), carbon dioxide (CO_2) or carbon monoxide (CO), and phosphate (PO_4^{3-}). The pre-life Earth environment was also plenteous in water (H_2O). There was little or no free oxygen (O_2) and ozone (O_3). In order for life to arise on Earth, organic molecules such as amino acids (the building blocks of proteins), sugars, acids, and bases would need to have formed from these available chemicals.

Over the last century there has been much research into plausible mechanisms for the origin of life. The **Oparin Hypothesis** is one theory developed by a Russian scientist (A.I. Oparin) in 1924. Oparin proposed that the Earth was formed approximately 4.6 billion years ago, stating that the early Earth had a reducing atmosphere, meaning there was very little free oxygen present. Instead there was an abundance of ammonia, hydrogen, methane, and steam (H_2O), all escaping from volcanoes. The Earth was in the process of cooling down, so there was a great deal of heat energy available, as well as a pattern of recurring violent lightning storms providing a source of energy in addition to sunlight. During this cooling of the Earth, much of the steam surrounding the Earth would condense, forming hot seas. In the presence of abundant energy, the synthesis of simple organic molecules from the available chemicals would be possible. These organic substances then collected in the hot, turbulent seas (sometimes referred to as the "primordial soup").

As the concentration of organic molecules became very high, they formed into larger, charged complex molecules. Oparin called these highly absorptive molecules "coacervates." Coacervates theoretically had the ability to divide.

Oparin's research involved experimental findings of evidence to support his ideas that amino acids could combine to form proteins under early Earth conditions. Oparin knew proteins were catalysts, so they could encourage further change and development of early cells.

Stanley Miller provided support for Oparin's hypotheses in experiments where he exposed simple inorganic molecules to electrical charges similar to lightning. Miller recreated conditions as they were supposed to have existed in early Earth history, and was successful in his attempts to produce complex organic molecules including amino acids under these conditions. Miller's experiments served to support Oparin's hypotheses.

Sidney Fox, a major evolution researcher of the 1960s, conducted experiments that proved ultraviolet light may induce the formation of dipeptides from amino acids. Under conditions of moderate dry heat, Fox showed formation of proteinoids—polypeptides of up to 18 amino acids. He also showed that poly-phosphoric acid could increase the yield of these polymers, a process that simulates the modern role of ATP in protein synthesis. These proteins formed small spheres known as microspheres that showed similarities to living cells.

Fig. 2-1 Setup of the Miller-Urey Experiment.

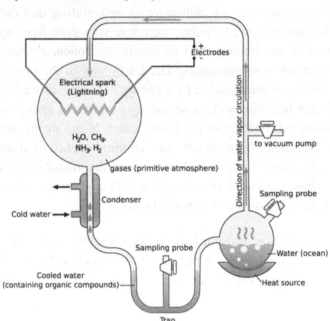

Further strides were made by researcher Cyril Ponnamperuma who demonstrated that small amounts of guanine formed from the thermal polymerization of amino acids. He also demonstrated the synthesis of adenine and ribose from long-term treatment of reducing atmospheric gases with electrical current.

Once organic compounds had been synthesized, it is theorized that primitive cells developed that used ATP for energy and contained genetic material in the form of RNA (or possibly DNA). These primitive cells, called prokaryotes, were similar to some bacteria now found on Earth.

The endosymbiont theory suggests that original prokaryotic cells absorbed other cells that performed various tasks. For instance, an original cell could absorb several symbiotic bacteria that then evolve into mitochondria. Several cells that lived in symbiosis would have combined and evolved to form a single eukaryotic cell.

Fig. 2-2 Endosymbiont Theory. Step (1) Two prokaryotic cells (A and B) combine to form a new cell that is more complex since the functions of the ingested cell can continue within the first. The new cell combines with C to form D and so on. Eventually they form a eukaryotic cell (I) with diverse membrane-bound organelles performing specified functions within that single cell.

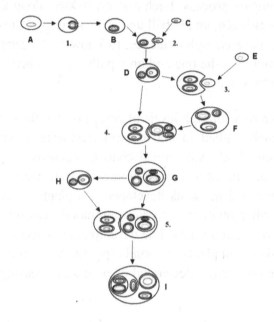

EVOLUTION OF LIFE

Plant Evolution

The evolution of plant species is still a matter of considerable research and discovery. Most theories indicate that the first plants derived from heterotrophic prokaryotic cells (cells that got their energy from other cells). Since it is presumed that the early Earth's atmosphere was lacking in oxygen, early cells were anaerobic. Over time, some bacteria evolved the ability to carry on photosynthesis (most likely the cyano-bacteria), thus becoming autotrophic (making their own food/energy), which in turn introduced significant amounts of oxygen into the atmosphere. Oxygen is poisonous to most anaerobic cells, however, a new niche opened up: cells became able to survive in the presence of oxygen, and were also able to use oxygen for metabolism.

Cyanobacteria were incorporated into larger aerobic cells, which then evolved into photosynthetic eukaryotic cells. Cellular organization increased as nuclei and membranes formed, and cell specialization occurred, leading to multicellular photosynthetic organisms—that is, plants.

Clearly, this is a very abbreviated and simplified version of a very prolonged and extremely complex process. Each and every step along the way has been researched and questioned, and is still under the scrutiny of research. Scientists simply do not yet have enough evidence, to know with certainty whether one pathway of evolution was the one and only pathway or whether another or possibly several pathways converged.

Scientists do know with high levels of certainty that the earliest plants were aquatic, but as niches filled in marine and freshwater environments, plants began to move onto land. Anatomical changes occurred over time, allowing plants to survive in a non-aqueous environment. Cell walls thickened and tissues to carry water and nutrients developed. As plants continued to adapt to land conditions, differentiation of tissues continued, resulting in the evolution of stems, leaves, roots, and seeds. The development of the seed was a key factor in the survival of land plants. Asexual reproduction dominated in early species, but sexual reproduction developed over time, increasing the possibilities of diversity.

Processes of adaptive radiation, genetic drift, and natural selection continued over long periods of time to produce the incredible diversity seen in the plant world today.

Animal Evolution

The evolution of animals is thought to have begun with marine protists (one-celled living organisms). Although there is no fossil record going back to the protist level, animal cells bear the most similarity to marine protist cells. Fossilized burrows from multicellular organisms begin to appear in the geological record approximately 700 million years ago, during the Precambrian period. These multicellular animals had only soft parts—no hard parts, which could be fossilized. We can only see the remains of their life—the burrows they made.

During the Cambrian period (the first period of the Paleozoic Era), beginning about 570 million years ago, the fossil record begins to show multicellular organisms with hard parts, exoskeletons. The fossil record at this time includes fossil representations from all modern day (and some extinct) phyla. This sudden appearance of multitudes of differentiated animal forms is known as the **Cambrian explosion.**

At the end of the Paleozoic Era, the fossil record attests to several mass extinction events. These combined events resulted in the extinction of about 95% of animal species developed to that point. Fossils indicate that many organisms, such as Trilobites, which were numerous in the Cambrian era, did not survive to the end of the Paleozoic Era.

Approximately 505 million years ago marked the beginning of the Ordovician period, which lasted until about 440 million years ago. The Ordovician period was marked by diversification among species that survived past the Cambrian extinctions. The Ordovician is also known for the development of land plants. Early forms of fish arose in the Cambrian, but developed during the Ordovician, these being the first vertebrates to be seen in the fossil record. Again, the end of the Ordovician is marked by vast extinctions, but these extinctions allowed the opening of ecological situations, which in turn encouraged adaptive radiation.

Adaptive radiation is the evolutionary mechanism (see section on evolutionary mechanisms) credited with the development of new species in the next period, the Silurian, from 440 to 410 million years ago. The Silurian period is marked by widespread colonization of landmasses by plants and animals. Large numbers of insect fossils are recognizable in Silurian geologic sediments, as well as fish and early amphibians. The mass movement onto land by formerly marine animals required adaptation in numerous areas, including gas exchange, support (skeletal), water conservation, circulatory systems, and reproduction.

In the study of animal evolution, attention is paid to two concepts, homology and analogy. Structures that exist in two different species because they share a common ancestry are called **homologous**. For instance, the forelimbs of a salamander and an opossum are similar in structure because of common ancestry. **Analogous** structures are similar because of their common function, although they do not share a common ancestry. Analogous structures are the product of **convergent evolution**. For instance, birds and insects both have wings, although they are not relatives. Rather, the wings evolved as a result of convergence. Convergence occurs when a particular characteristic evolves in two unrelated populations. Wings of insects and birds are analogous structures (they are similar in function regardless of the lack of common ancestors).

The process of **extinction** has played a large part in the direction evolution has taken. Extinctions occur at a generally low rate at all times. It is presumed that species that face extinction have not been able to adapt appropriately to environmental changes. However, there have also been several "extinction events" that have wiped out up to 95% of the species of their time. These events served to open up massive ecological niches, encouraging evolution of multitudes of new species.

Approximately 400 million years ago, the first amphibians gave rise to early reptiles that then diversified into birds, then mammals. One branch of mammals developed into the tree-dwelling primates, considered the ancestors of humans.

Human Evolution

Humans evolution theories are still in development, but current research states that *Homo sapiens sapiens* (modern humans) have evolved from primates who over time developed larger brains. A branch of bipedal primates gave rise to the first true hominids about 4.5 million years ago. The earliest known hominid fossils were found in Africa in the 1970s. The well-known "Lucy" skeleton was named *Australopithecus afarensis*. It was determined from the skeleton of *Australopithecus* that it was a biped. It had a human-like jaw and teeth, but a skull that was more similar to that of a small ape. The arms were proportionately longer than humans, indicating the ability to still be motile in trees.

The fossilized skull of *Homo erectus,* who is considered the oldest known fossil of the human genus, is thought to be about 1.8 million years old. The skull of *Homo erectus* was quite a lot larger than *Australopithecus,* about the size of a modern human brain. *Homo erectus* was thought to walk upright and had facial features more closely resembling humans than apes. The oldest fossils to be designated *Homo sapiens* are also called Cro-Magnon man, with brain size and facial features essentially the same as modern humans. Cro-Magnon *Homo sapiens* are thought to have evolved in Africa and migrated to Europe and Asia approximately 100,000 years ago.

Fig. 2-3 Geologic Time Scale.

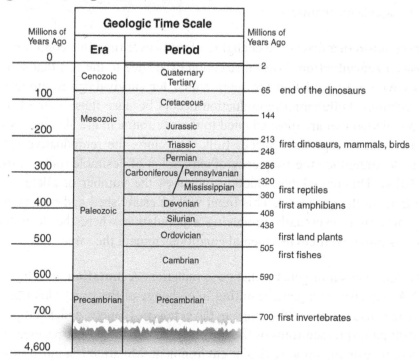

MECHANISMS OF EVOLUTION

Modern understanding of the process of natural selection recognizes that there are some basic mechanisms that support evolutionary change. Mutations happen only in individuals, and Darwin focused on natural selection that happened to individuals. However, modern theories focus on the change that occurs among populations within communities, not on individuals. The mutations that occur produce traits that affect individuals that then pass down traits through their genes. This then has an impact on a population of individuals after several generations. Natural selection that impacts evolution is the selection of genetic traits that make a population of a species more fit to survive and to reproduce in the community.

All evolution is dependent upon genetic change. The entire collection of genes within a given population is known as its **gene pool**. Individuals in the population will have only one pair of alleles (variations of a gene type) for a particular single-gene trait. Yet, the gene pool may contain dozens or hundreds of alleles for this trait. Evolution does not occur through changes from individual to individual, but rather as the gene pool changes through one of a number of possible mechanisms.

One mechanism that drives the changing of traits over time in a community is **differential reproduction**. Natural selection exists due to the fact that some individuals within a population are more suited for survival, given environmental conditions. Differential reproduction occurs because those individuals within a population that are most adapted to the environment are also the most likely individuals to reproduce successfully. Therefore, the reproductive processes tend to strengthen the frequency of expression of desirable traits across the population. Differential reproduction increases the number of alleles for desirable traits in the gene pool. This trend will be established and strengthen gradually over time, eventually producing a population where the desirable trait becomes rampant, if environmental conditions remain the same.

Another mechanism of genetic change is mutation. A **mutation** is a change of the DNA sequence of a gene, resulting in a change of the trait. Although a mutation can cause a very swift change in the genotype (genetic code) and possibly phenotype (expressed trait) of the offspring, mutations do not necessarily produce a trait desirable for a particular environment. Mutation is a much more random occurrence than differential reproduction.

Although mutations occur quickly, the change in the gene pool is minimal, so change in the community occurs very slowly (over multiple generations). Mutation does provide a vehicle of introducing new genetic possibilities; genetic traits, which did not exist in the original gene pool, can be introduced through mutation.

A third mechanism recognized to influence the evolution of new traits is known as **genetic drift**. Over time, a gene pool (particularly in a small population) may experience a change in frequency of particular genes simply due to chance fluctuations. In a finite population, the gene pool may not reflect the entire number of genetic possibilities of the larger genetic pool of the species population. Over time, the genetic pool within this finite population changes, and evolution has occurred. Genetic drift has no particular tie to environmental conditions, and thus the random change in gene frequency is unpredictable. The change of gene frequency may produce a small or a large change, depending on what traits are affected. The process of genetic drift, as opposed to mutation, actually causes a reduction in genetic variety.

Genetic drift occurs within finite separated populations, allowing each population to develop its own distinct gene pool. However, occasionally an individual from an adjacent population of the same species may immigrate and breed with a member of the previously locally isolated group. The introduction of new genes from the immigrant results in a change of the gene pool, known as **gene migration**. Gene migration is also occasionally successful between members of different, but related, species. The resultant hybrids succeed in adding increased variability to the gene pool.

The study of genetics shows that in a situation where random mating is occurring within a population (which is in equilibrium with its environment), gene frequencies and genotype ratios will remain constant from generation to generation. This law is known as the **Hardy-Weinberg Law of Equilibrium**, named after the two men (G.H. Hardy and Wilhelm Weinberg, c. 1909) who first studied this principle in mathematical studies of genetics. The Hardy-Weinberg Law is a mathematical formula that shows why recessive genes do not disappear over time from a population.

According to the Hardy-Weinberg Law, the sum of the frequencies of all possible alleles for a particular trait is 1. That is,

$$p + q = 1.0$$

where the frequency of one allele is represented by **p** and the frequency of another is **q**. It then follows mathematically that the frequency of genotypes within a population can be represented by the equation:

$$p^2 + 2pq + q^2 = 1$$

where the frequency of homozygous dominant genotypes is represented by p^2, the homozygous recessive by q^2, and the heterozygous genotype by $2pq$.

For instance, in humans the ability to taste the chemical phenylthiocarbamide (PTC) is a dominant inherited trait. If **T** represents the allele for tasting PTC and **t** represents the recessive trait (inability to taste PTC), then the possible genotypes in a population would be **TT, Tt,** and **tt**. If the frequency of non-tasters in a particular population is 4% or 0.04 (that is, $q^2 = 0.04$), then the frequency of the allele **t** equals the square root of 0.04 or 0.2. It is then possible to calculate the frequency of the dominant allele, **T**, in the population using the equation:

$$p + 0.2 = 1 \text{ so, } p = 0.8$$

The frequency of the allele for tasting PTC is 0.8. The frequency of the various possible genotypes (**TT, Tt,** and **tt**) in the population can also be calculated since the frequency of the homozygous dominant is p^2 or 0.64 or 64%. The frequency of the heterozygous genotype is $2pq$.

$$2pq = 2(0.8)(0.2) = 0.32 = 32\%.$$

$$\text{Frequency of TT} = 64\%, \text{Tt} = 32\%, \text{tt is } 4\% \ldots \text{totaling } 100\%$$
$$\text{or } 0.64 + 0.32 + 0.04 = 1$$

In order for Hardy-Weinberg equilibrium to occur, the population in question must meet the following conditions: random mating (no differential reproduction) must be taking place and no migration, mutation, selection, or genetic drift can be occurring. When these conditions are met, Hardy-Weinberg equilibrium can occur, and there will be no changes in the gene pool over time. Hardy-Weinberg is important to the evolutionary process because it shows that alleles that have no current selective value will be retained in a population over time.

MECHANISMS OF SPECIATION

A species is an interbreeding population that shares a common gene pool and produces viable offspring. Up to this point we have been considering mechanisms that produce variation within species. It is apparent that to explain evolution on a broad scale we must understand how genetic change produces new species. There are two mechanisms that produce separate species: allopatric speciation and sympatric speciation.

In order for a new species to develop, substantial genetic changes must occur between populations, which prohibit them from interbreeding. These genetic changes may result from genetic drift or from mutation that take place separately in the two populations. **Allopatric speciation** occurs when two populations are geographically isolated from each other. For instance, a population of squirrels may be geographically separated by a catastrophic event such as a volcanic eruption. Two populations (separated by the volcanic flow) continue to reproduce and experience genetic drift and/or mutation over time. This limits each population's gene pool and produces changes in expressed traits. Later, the geographical separation may be eliminated as the volcanic flow subsides; even so, the two populations have now experienced too much change to allow them to successfully interbreed again. The result is the production of two separate species.

Speciation may also occur without a geographic separation when a population develops members with a genetic difference, which prevents successful reproduction with the original species. The genetically different members reproduce with each other, producing a population, which is separate from the original species. This process is called **sympatric speciation**.

As populations of an organism in a given area grow, some will move into new geographic areas looking for new resources or to escape predators. (In this case, a natural event does not separate the population; instead, part of the population moves.) Some of these adventurers will discover new niches and advantageous conditions. Traits possessed by this traveling population will grow more common over several generations through the process of natural selection. Over time the species will specially adapt to live more effectively in the new environment. Through this process, known as **adaptive radiation**, a single species can develop into several diverse species over time. If the separated populations merge again and are able to successfully interbreed, then by definition new species have not been developed. Adaptive radiation is proven to have occurred when the species remerge and do not interbreed successfully.

All of the above evolutionary mechanisms are dependent upon reproduction of organisms over a long period of time, a very gradual process. **Punctuated equilibrium** is an entirely different method of explaining speciation. Punctuated equilibrium is a scientific model that proposes that adaptations of species arise suddenly and rapidly. Punctuated equilibrium states that species undergo a long period of equilibrium, which at some point is upset by environmental forces causing a short period of quick mutation and change.

Punctuated equilibrium was first proposed as paleontologists studied the fossil record. Gradualism would produce slowly changing and adapting species over many generations. However, the fossil record seems to show that organisms in general survive many generations in many areas with very little change over long periods of geologic time. New species appear in the fossils suddenly, without transitional forms, though "sudden" in this context needs to be understood on a geologic time scale.

Scientists still do not agree on the degree to which gradualism, punctuated equilibrium, or a combination of these processes is responsible for speciation.

CLASSIFICATION OF LIVING ORGANISMS

The study of **taxonomy** seeks to organize living things into groups based on morphology, or more recently, genetics.

Carolus Linnaeus, who published his book *Systema Naturae* in 1735, first developed our current methods of taxonomy. Linnaeus based his taxonomic keys on the morphological (outward anatomical) differences seen among species. Linnaeus designed a system of classification for all known and unknown organisms according to their anatomical similarities and differences.

Linnaeus used two Latin-based categories—*genus* and *species*—to name each organism. Every genus name could include one or more types of species. We refer to this two-word naming of species as **binomial nomenclature** (literally meaning "two names" in Latin). For example, Linnaeus named humans *Homo sapiens* (literally "man who is wise"). *Homo* is the genus name and *sapiens* the species name. *Homo sapiens* is the only extant species left from the genus *Homo*.

Fig. 2-4   Organism Classification System.

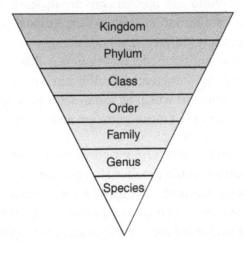

Beyond genus and species, Linnaeus further categorized organisms in a total of seven levels. Every **species** also belongs to a **genus, family, order, class, phylum,** and **kingdom**. *Kingdom* is the most general category, *species* the most limited. Taxonomists now also add "sub" and "super" categories to give even more opportunity for grouping similar organisms, and have added categories even more general than kingdom (that is, **domains**).

When considering the classification of living organisms, and as you look at Figure 2-4, you should realize that the organisms within the classification layers as you go down the upside-down pyramid are going to be more and more alike to each other. For example, organisms within the same kingdom will be more alike than those in another kingdom, but organisms in the same genus will be much more alike than those sharing only a family, but not a genus.

Although Linnaeus's system was designed before evolution was understood, the system he designed remains in use because it is based on similarities of species. An evolutionary classification system would use these same criteria. However, today the classification system also serves to show relationships between organisms. When one constructs a **phylogenetic tree** (an evolutionary family tree of species) the result will generally show the same relationships represented with Linnaeus's taxonomy.

The most modern classification system contains three domains: the **Archaea**, the **Eubacteria**, and the **Eukaryota**. The organisms of the domain Archaea are prokaryotic, have unique RNA, and are able to live in the extreme ecosystems on Earth. The domain Archaea includes methane-producing organisms, and organisms able to withstand extreme temperatures and high salinity. The domain Eubacteria contains the prokaryotic organisms we call bacteria.

It should be noted that some taxonomists still use a system including five kingdoms and no domains. In this case, the kingdom Monera would include organisms that are considered by other taxonomists to be included within the domains Archaea and Eubacteria.

The domain Eukaryota includes all organisms that possess eukaryotic cells. The domain Eukaryota includes the four kingdoms: **Kingdom Protista, Kingdom Fungi, Kingdom Animalia,** and **Kingdom Plantae**. The following chart gives the major features of the four kingdoms of the Eukaryota:

Fig. 2-5 Four Kingdoms of the Eukaryota.

Kingdom	No. of Known Phyla/Species	Nutrition	Structure	Included Organisms
Protista	27/250,000 +	photosynthesis, some ingestion and absorption	large eukaryotic cells	algae & protozoa
Fungi	5/100,000 +	absorption	multicellular (eukaryotic) filaments	mold, mushrooms, yeast, smuts, mildew
Animalia	33/1,000,000 +	ingestion	multicellular, specialized eukaryotic motile cells	various worms, sponges, fish, insects, reptiles, amphibians, birds, and mammals
Plantae	10/250,000 +	photosynthesis	multicellular, specialized eukaryotic nonmotile cells	ferns, mosses, woody and non-woody flowering plants

There are nine major phyla within the **Kingdom Animalia**. (Each phylum is further broken down, but the focus here will be on the phylum Chordata.) The Phyla are as follows:

1. **Porifera**—the sponges
2. **Cnidaria**—jellyfish, sea anemones, hydra, etc.
3. **Platyhelminthes**—flat worms
4. **Nematoda**—round worms
5. **Mollusca**—snails, clams, squid, etc.
6. **Annelida**—segmented worms (earthworms, leeches, etc.)

7. **Arthropoda**—crabs, spiders, lobster, millipedes, insects
8. **Echinodermata**—sea stars, sand dollars, etc.
9. **Chordata**—fish, amphibians, reptiles, birds, mammals, lampreys

Vertebrates are within the phylum Chordata, which is split into three subphyla, the **Urochordata** (animals with a tail cord such as tunicates), the **Cephalochordata** (animals with a head cord such as lampreys), and **Vertebrata** (animals with a backbone).

The subphylum Vertebrata is divided into two **superclasses**, the **Aganatha** (animals with no jaws), and the **Gnathostomata** (animals with jaws). The Gnathostomata includes six classes with the following major characteristics:

a. **Chondrichthyes**—fish with a cartilaginous endoskeleton, two-chambered heart, 5–7 gill pairs, no swim bladder or lung, and internal fertilization (sharks, rays, etc.).

b. **Osteichthyes**—fish with a bony skeleton, numerous vertebrae, swim bladder (usually), two-chambered heart, gills with bony gill arches, and external fertilization (herring, carp, tuna).

c. **Amphibia**—animals with a bony skeleton, usually with four limbs having webbed feet with four toes, cold-blooded (ectothermic), large mouth with small teeth, three-chambered heart, separate sexes, internal or external fertilization, amniotic egg (salamanders, frogs, etc.).

d. **Reptilia**—horny epidermal scales, usually have paired limbs with five toes (except limbless snakes), bony skeleton, lungs, no gills, most have three-chambered heart, cold-blooded (ecothermic), internal fertilization, separate sexes, mostly egg-laying (oviparous), eggs contain extraembryonic membranes (snakes, lizards, alligators).

e. **Aves**—spindle-shaped body (with head neck, trunk, and tail), long neck, paired limbs, most have wings for flying, four-toed foot, feathers, leg scales, bony skeleton, bones with air cavities, beak, no teeth, four-chambered heart, warm blooded (endothermic), lungs with thin air sacs, separate sexes, egg-laying, eggs have hard calcified shell (birds, ducks, sparrows, etc.).

f. **Mammalia**—body covered with hair, glands (sweat, scent, sebaceous, mammary), teeth, fleshy external ears, usually four limbs, four-chambered heart, lungs, larynx, highly developed brain, warm-blooded, internal fertilization, live birth (except for the egg-laying monotremes), milk-producing (cows, humans, platypus, apes, etc.).

CHAPTER 3

Cellular and Molecular Biology

CHAPTER 3

CELLULAR AND MOLECULAR BIOLOGY

THE STRUCTURE AND FUNCTION OF CELLS

The **cell** is the smallest and most basic unit of most living things **(organisms)**. Many species have only a single cell, others are multicellular. (Although viruses are sometimes considered to be living, they are non-cellular and cannot fulfill the characteristics of life without invading the cell of another organism.) Cell structure varies according to the function of the cell and the type of living thing.

There are two main types of cells: prokaryotic and eukaryotic. **Prokaryotes** have no nucleus or any other membrane-bound **organelles** (cell components that perform particular functions). The DNA in prokaryotic cells usually forms a single chromosome which is circular or loop-like and which floats within the cytoplasm. Prokaryotic organisms have only one cell and include all bacteria. Plant, fungi, and animal cells, as well as protozoa, are **eukaryotic**. Eukaryotic cells contain membrane-bound intracellular organelles, including a nucleus. The DNA within eukaryotes is organized into chromosomes.

Fig. 3-1 Comparison of Eukaryote and Prokaryote Cells.

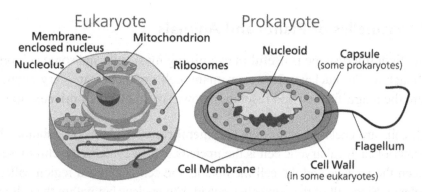

A single organism can be *unicellular* (consisting of just one cell), or *multicellular* (consisting of many cells). A multicellular organism may have many different types of cells that differ in structure to serve different functions. Individual cells may contain organelles that assist them with specialized functions. For example, muscle cells tend to contain more mitochondria (organelles that make energy available to the cells) since muscle requires the use of extra energy.

Animal cells differ in structure and function from photosynthetic cells, which are found in plants and some protists. Photosynthetic cells have the added job of producing food so they are equipped with specialized photosynthetic organelles. Plant cells also have a central vacuole and cell walls, structures not found in animal cells.

Note: Some prokaryotes are photosynthetic as well, but don't "contain" photosynthetic cells. They simply have developed photosynthetic ability (i.e., cyanobacteria).

Fig. 3-2 Cyanobacteria (Photosynthetic Prokaryote).

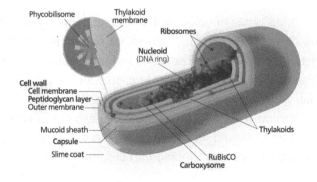

Cell Organelles of Plants and Animals

The light microscope is useful in examining most cells and some cell organelles (such as the nucleus). However, many cell organelles are very small and require the magnification and resolution power of an **electron microscope**.

All cells are enclosed within the **cell membrane** (or plasma membrane). Near the center of each eukaryotic cell is the **nucleus**, which contains the chromosomes. Between the nucleus and the cell membrane, the cell contains a region called the **cytoplasm**. Since all of the organelles outside the nucleus but within the cell membrane exist within the cytoplasm, they are all called **cytoplasmic organelles**.

The shape and size of cells can vary widely. The longest nerve cells (neurons) may extend over a meter in length with an approximate diameter of only 4-100 micrometers (1 millimeter = 1,000 micrometers [μm]). A human egg cell may be 100 micrometers in diameter. The average size of a bacterium is 0.5 to 2.0 micrometers. However, most cells are between 0.5 and 100 micrometers in diameter. The size of a cell is limited by the ratio of its volume to its surface area. In the illustration below, note the variation of shape of cells within the human body:

Fig. 3-3 Varying Cell Types. The six sketches of human cell types show some of the diversity in shape and size among cells with varying functions. The sketches are not sized to scale.

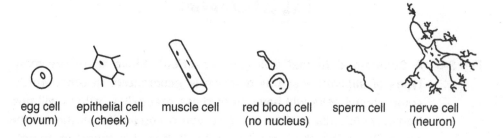

egg cell (ovum)   epithelial cell (cheek)   muscle cell   red blood cell (no nucleus)   sperm cell   nerve cell (neuron)

Animal Cells (See Fig. 3-5.)

1. The **cell membrane (1a)** encloses the cell and separates it from the environment. It may also be called a plasma membrane. This membrane is composed of a double layer (bilayer) of phospholipids with globular proteins embedded within the layers. These proteins span all layers of the membrane connecting the inside of the cell with the outside world. The membrane is extremely thin (about 80 angstroms; 10 million angstroms = 1 millimeter) and elastic. The combination of the lipid bilayer and the proteins embedded within it allow the cell to determine what molecules and ions can enter and leave the cell, and regulate the rate at which they enter and leave.

Endocytic vesicles (1b) form when the plasma membrane of a cell surrounds a molecule outside the membrane, then releases a membrane-bound sack containing the desired molecule or substance into the cytoplasm. This process allows the cell to absorb larger molecules than would be able to pass through the cell membrane, or that need to remain packaged within the cell.

2. **Microvilli** are projections of the cell extending from the cell membrane. Microvilli are found in certain types of cells—for example, those involved in absorption (such as the cells lining the intestine). These filaments increase the surface area of the cell membrane, increasing the area available to absorb nutrients. They also contain enzymes involved in digesting certain types of nutrients.

Fig. 3-4 **Cell Membrane. A phospholipid bilayer with embedded globular proteins.**

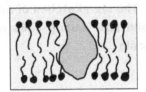

Fig. 3-5 **A Generalized Animal Cell (cross-section). Since there are many types of animal cells, this diagram is generalized. In other words, some animal cells will have all of these organelles, others will not. However, this illustration will give you a composite picture of the organelles within the typical animal cell. It is also important to note the function of each organelle. Each labeled component is explained by the corresponding text.**

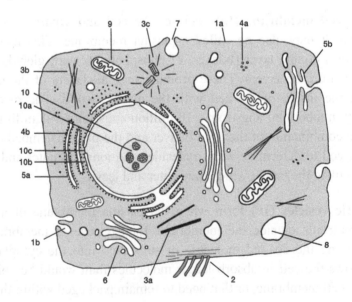

3. The **cytoskeleton** provides structural support to a cell. **Microtubules (3a)** are long, hollow, cylindrical protein filaments, which give structure to the cell. These filaments are scattered around the edges of a cell and form a sort of loose skeleton or framework for the cytoplasm. Microtubules also are found at the base of cilia or flagella (organelles which allow some cells to move on their own) and give these organelles the ability to move. **Microfilaments (3b)** are double-stranded chains of proteins, which serve to give structure to the cell. Together with the larger microtubules, microfilaments form the cytoskeleton, providing stability and structure. **Centrioles (3c)** are structural components of many cells, and are particularly common in animal cells. Centrioles are tubes constructed of a geometrical arrangement of microtubules in a pinwheel shape. Their function includes the formation of new microtubules, but is primarily the formation of structural skeleton around which cells split during mitosis and meiosis. Basal bodies are structurally similar to centrioles, but their function is to anchor and aid in the movement of flagella or cilia.

Fig. 3-6 Cross-section of a Centriole. Centrioles are tubes constructed of a geometrical arrangement of microtubules in a pinwheel shape.

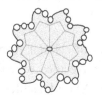

4. **Ribosomes** are the site of protein synthesis within cells. Ribosomes are composed of certain protein molecules and RNA molecules (ribosomal RNAs, or rRNAs). **Free ribosomes (4a)** float unattached within the cytoplasm. The proteins synthesized by free ribosomes are made for use in the cytoplasm, not within membrane-bound organelles. **Attached ribosomes (4b)** are attached to the ER (see No. 5). Proteins made at the site of attached ribosomes are destined for use within the membrane-bound organelles.

5. The **endoplasmic reticulum,** a large organization of folded membranes, is responsible for the delivery of lipids and proteins to certain areas within the cytoplasm (a sort of cellular highway). **Rough endoplasmic reticulum** or **RER (5a)** has attached ribosomes. In addition to packaging and transport of materials within the cell, the RER is instrumental to protein synthesis. **Smooth endoplasmic reticulum** or **SER (5b)** is a network of membranous channels. Smooth endoplasmic reticulum does not have attached ribosomes. The endoplasmic reticulum is

responsible for processing lipids, fats, and steroids, which are then packaged and dispersed by the Golgi apparatus.

6. The **Golgi apparatus** (also known as Golgi bodies, or the Golgi complex) is instrumental in the storing, packaging, and shipping of proteins. The Golgi apparatus looks much like stacks of hollow pancakes and is constructed of folded membranes. Within these membranes, cellular products are stored or packaged by closing off a bubble of membrane with the proteins or lipids inside. These packages are shipped (via the endoplasmic reticulum) to the part of the cell where they will be used or to the cell membrane for secretion from the cell.

7. **Secretory vesicles** are packets of material packaged by either the Golgi apparatus or the endoplasmic reticulum. Secretory vesicles carry substances produced within the cell (a protein, for example) to the cell membrane. The vesicle membrane fuses with the cell membrane in a process called **exocytosis**, allowing the substance to escape the cell.

8. **Lysosomes** are membrane-bound organelles containing digestive enzymes. Lysosomes digest unused material within the cell, damaged organelles, or materials absorbed by the cell for use.

9. **Mitochondria** are centers of cellular respiration (the process of breaking up covalent bonds within sugar molecules with the intake of oxygen and release of ATP, adenosine tri-phosphate). ATP molecules store energy that is later used in cell processes. Mitochondria (plural of mitochondrion) are more numerous in cells requiring more energy (muscle, etc.). Mitochondria are self-replicating, containing their own DNA, RNA, and ribosomes. Mitochondria have a double membrane; the internal membrane is folded. Cellular respiration reactions occur along the folds of the internal membrane (called **cristae**). Mitochondria are thought to be an evolved form of primitive bacteria (prokaryotic cells) that lived in a symbiotic relationship with eukaryotic cells more than 2 billion years ago. This concept, known as the **endosymbiont hypothesis**, is a plausible explanation of how mitochondria, which have many of the necessary components for life on their own, became an integral part of eukaryotic cells.

10. The **nucleus** is an organelle surrounded by two lipid bilayer membranes. The nucleus contains chromosomes, nuclear pores, nucleoplasm, and nucleoli. The **nucleolus (10a)** is a rounded area within the nucleus of the cell where ribosomal RNA is synthesized. This rRNA is incorporated into ribosomes after exiting the nucleus. Several nucleoli (plural of nucleolus) can exist within a nucleus. The

nuclear membrane (10b) is the boundary between the nucleus and the cytoplasm. The nuclear membrane is actually a double membrane, which allows for the entrance and exit of certain molecules through the nuclear pores. **Nuclear pores (10c)** are points at which the double nuclear membrane fuses together, forming a passageway between the inside of the nucleus and the cytoplasm outside the nucleus. Nuclear pores allow the cell to selectively move molecules in and out of the nucleus. There are many pores scattered about the surface of the nuclear membrane.

Plant Cells

The structure of plant cells differs noticeably from animal cells with the addition of three organelles: the cell wall, the chloroplasts, and the central vacuole. In Fig. 3-7, the organelles numbered 1 to 7 function the same way in plant cells as in animal cells (see below).

Fig. 3-7 A Typical Plant Cell.

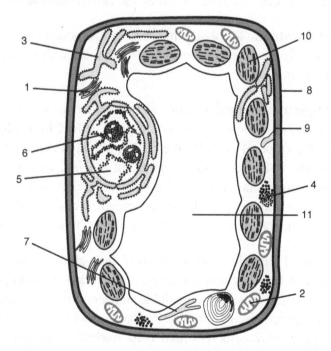

1. Golgi apparatus

2. Mitochondria

3. Rough endoplasmic reticulum

4. **Ribosome**

5. **Nucleus**

6. **Nucleolus**

7. **Smooth endoplasmic reticulum**

8. **Cell walls** surround plant cells. (Bacteria also have cell walls.) Cell walls are made up of cellulose and lignin, making them strong and rigid (whereas the cell membrane is relatively weak and flexible). The cell wall encloses the cell membrane providing strength and protection for the cell. The cell wall allows plant cells to store water under relatively high concentration. The combined strength of a plant's cell walls provides support for the whole organism. Dry wood and cork are essentially the cell walls of dead plants. The structure of the cell wall allows substances to pass through it readily, so transport in and out of the cell is still regulated by the cell membrane.

9. The **cell membrane** (or plasma membrane) functions in plant and animal cells in the same way. However, in some plant tissues, channels connect the cytoplasm of adjacent cells.

10. **Chloroplasts** are found in plant cells (and also in some protists). Chloroplasts are the site of photosynthesis within plant cells. **Chlorophyll** pigment molecules give the chloroplast their green color, although the chloroplasts also contain yellow and red carotenoid pigments. In the fall, as chloroplasts lose chlorophyll, these pigments are revealed giving leaves their red and yellow colors. The body (or **stroma**) of the chloroplast contains embedded stacked disk-like plates (called **grana**), which are the site of photosynthetic reactions.

11. The **central vacuole** takes up much of the volume of plant cells. It is a membrane-bound (this particular membrane is called the **tonoplast**), fluid-filled space, which stores water and soluble nutrients for the plant's use. The tendency of the central vacuole to absorb water provides for the rigid shape (turgidity) of some plant cells. (Animal cells may also contain vacuoles for varying purposes, and these too are membrane-bound fluid-filled spaces. For instance, contractile vacuoles perform the specific function of expelling waste and excess water from single-celled organisms.)

Properties of Cell Membranes

The cell membrane is an especially important cell organelle with a unique structure, which allows it to control movement of substances into and out of the cell. Made up of a fluid phospholipid bilayer, proteins, and carbohydrates, this extremely thin (approximately 80 angstroms) membrane can only be seen clearly with an electron microscope. The selective permeability of the cell membrane serves to manage the concentration of substances within the cell. Substances can cross the cell membrane by passive transport, facilitated diffusion, and active transport. During **passive transport**, substances freely pass across the membrane without the cell expending any energy. **Facilitated diffusion** does not require added energy, but it cannot occur without the help of specialized proteins. Transport requiring energy output from the cell is called **active transport**.

Simple diffusion is one type of passive transport. **Diffusion** is the process whereby molecules and ions flow through the cell membrane from an area of higher concentration to an area of lower concentration (thus tending to equalize concentrations). Where the substance exists in higher concentration, collisions occur, which tend to propel them away toward lower concentrations. Diffusion generally is the means of transport for ions and molecules that can slip between the lipid molecules of the membrane. Diffusion requires no added energy to propel substances through a membrane.

Fig. 3-8 Diffusion. CO_2 diffuses out of the cell since its concentration is higher inside the cell. O_2 diffuses into the cell because its concentration is higher outside. Molecules diffuse from areas of high concentration to areas of lower concentration.

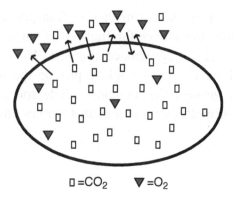

Another type of passive transport is **osmosis**, a special process of diffusion occurring only with water molecules. Osmosis does not require the addition of any energy, but occurs when the water concentration inside the cell differs from the concentration outside the cell. The water on the side of the membrane with the highest water concentration will move through the membrane until the concentration is equalized on both sides. When the water concentration is equal inside and outside the cell, it is called isomotic or isotonic. For instance, a cell placed in a salty solution will tend to lose water until the solution outside the cell has the same concentration of water molecules as the cytoplasm (the solution inside the cell).

Fig. 3-9 Osmosis. Water crosses the membrane into a cell that has a higher concentration of sugar molecules than the surrounding solution. Water crosses the membrane to leave the cell when there is a higher concentration of Na+ ions outside the cell than inside the cell.

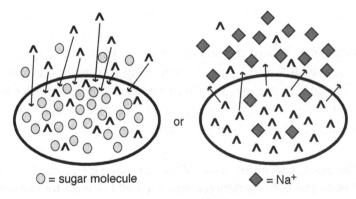

○ = sugar molecule ◆ = Na+
∧ = water molecule

Facilitated diffusion is another method of transport across the cell membrane. Facilitated diffusion allows for transfer of substances across the cell membrane with the help of specialized proteins. These proteins, which are embedded in the cell membrane, are able to pick up specific molecules or ions and transport them through the membrane. The special protein molecules allow the diffusion of molecules and ions that cannot otherwise pass through the lipid bilayer.

Fig. 3-10 Facilitated Diffusion. Specialized proteins embedded in the cell membrane (and crossing the entire membrane) permit passage of substances of a particular shape and size.

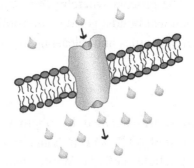

Active transport, like facilitated diffusion, requires membrane-bound proteins. Unlike facilitated diffusion, active transport uses energy to move molecules across a cell membrane against a concentration gradient (in the opposite direction than they would go under normal diffusion circumstances). With the addition of the energy obtained from ATP, a protein molecule embedded in the membrane changes shape and moves a molecule across the membrane against the concentration gradient.

Large molecules are not able to pass through the cell membrane, but may be engulfed by the cell membrane. **Endocytosis** is the process whereby large molecules (i.e., some sugars or proteins) are taken up into a pocket of membrane. The pocket pinches off, delivering the molecules, still inside a membrane sack, into the cytoplasm. This process, for instance, is used by white blood cells to engulf bacteria. **Exocytosis** is the reverse process, exporting substances from the cell.

Energy Transformations (Bioenergetics)

All living things require energy. Ultimately, the source of most energy for life on Earth is the sun. Photosynthetic organisms (plants, some protists, and some bacteria) are able to harvest solar energy and transform it into chemical energy eventually stored within covalent bonds of molecules (such as carbohydrates, fats, and proteins). These organisms are called primary producers. Consumers eat producers and utilize the chemical energy stored in them to carry on the functions of life. Other organisms then consume the consumers. In each of these steps along the food chain, some energy is lost as heat (see discussion of thermodynamic laws in Chapter 8).

Cellular metabolism is a general term that includes all types of energy transformation processes, including photosynthesis, respiration, growth, movement, etc. Energy transformations occur as chemicals are broken apart or synthesized within the cell. The process whereby cells build molecules and store energy (in the form of chemical bonds) is called **anabolism**. **Catabolism** is the process of breaking down molecules and releasing stored energy.

ATP

Energy from the sun is transformed by photosynthetic organisms into chemical energy in the form of ATP. **ATP (adenosine triphosphate)** is known as the energy currency of cellular activity. While energy is stored in the form of carbohydrates, fats, and proteins, the amount of energy contained within the bonds of any of these substances would overwhelm (and thus kill) a cell if released at once. In order for the energy to be released in small packets usable to a cell, large molecules need to be broken down in steps. ATP is an efficient storage molecule for the energy needed for cellular processes. ATP consists of a nitrogenous base (adenine), a simple sugar (ribose), and three phosphate groups. When a cellular process requires energy, a molecule of ATP can be broken down into ADP (adenosine diphosphate) plus a phosphate group. Even more energy is released when ATP is decomposed into AMP (adenosine monophosphate) and two phosphate groups. These energy-releasing reactions are then coupled with energy-absorbing reactions.

Photosynthesis

The process of **photosynthesis** includes a crucial set of reactions. These reactions convert the light energy of the sun into chemical energy usable by living things. Photosynthetic organisms carry out photosynthesis. They use the converted energy for their own life processes, and also store energy that may be used by organisms that consume them.

Although the process of photosynthesis actually occurs through many small steps, the entire process can be summed up with the following equation:

$$6CO_2 + 6H_2O + \text{light energy} \rightarrow C_6H_{12}O_6 + 6O_2$$

(carbon dioxide + water → glucose + oxygen)

Chlorophyll is a green pigment (a pigment is a substance that absorbs light energy). Photosynthesis occurs in the presence of chlorophyll, as the chlorophyll is able to absorb a photon of light. Chlorophyll is contained in the grana of the chloroplast (see discussion of plant cells earlier in this chapter). Photosynthesis can only occur where chlorophyll is present. It is not used up in the photosynthetic process, but must be present for the reactions to occur.

There are two phases of the photosynthetic process, the light reaction, or **photolysis**, and the dark reaction, or **CO_2 fixation.** During photolysis, the chlorophyll pigment absorbs a photon of light, leaving the chlorophyll in an excited (higher energy) state. The light reaction is a decomposition reaction, which separates water molecules into hydrogen and oxygen atoms utilizing the energy from the excited chlorophyll pigment. Oxygen, which is not needed by the cell, combines to form O_2 (gas) and is released into the environment. The free hydrogen is grabbed and held by a particular molecule (called the hydrogen acceptor) until it is needed. The excited chlorophyll also supplies energy to a series of reactions that produce ATP from ADP and inorganic phosphate (Pi).

The dark reaction (CO_2 fixation) then occurs in the stroma of the chloroplast. This second phase of photosynthesis does not require light; however, it does require the use of the products of photolysis (hydrogen and energy in the form of ATP). In this phase, six CO_2 molecules are linked with hydrogen (produced in photolysis) forming glucose (a six-carbon sugar). This is a multi-step process, which requires energy, the ATP produced in the photolysis phase. Glucose molecules can link to form polysaccharides (starch or sugar), which are then stored in the cell.

Cellular Respiration

Unlike photosynthesis (which only occurs in photosynthetic cells), respiration occurs in all cells. Respiration is the process that releases energy for use by the cell. There are several steps involved in cellular respiration. Some require oxygen (that is, they are **aerobic**) and some do not (that is, they are **anaerobic** reactions).

Glycolysis is the breaking down of the six-carbon sugar (glucose) into smaller carbon-containing molecules yielding ATP (glyco = sugar, lysis = breakdown). It is the first step in all respiration pathways and occurs in the cytoplasm of all living cells. Each molecule of glucose (six carbons) is broken down into two molecules of pyruvic acid (or pyruvate with three carbons each), two ATP molecules, and two hydrogen atoms (attached to NADH, nicotinamide adenine dinucleotide). This is an **anaerobic reaction** (no oxygen is required). After glycolysis has occurred, respiration will continue on one of two pathways, depending upon whether oxygen is present or not.

The process of cellular respiration is summarized by the following chemical equation:

$$C_6H_{12}O_6 + 6O_2 \rightarrow 6CO_2 + 6H_2O + ATP$$

(glucose + oxygen → carbon dioxide + water + energy)

CHEMICAL NATURE OF THE GENE

Watson and Crick were responsible for explaining the structure of the DNA molecule, research that laid the foundation of our current understanding of the function of chromosomes and genes. Today, through the discoveries of these two scientists, and through the collaborative work of scientists worldwide, the study of chromosomes and genetic inheritance has proceeded to discover the intricacies of the **genomes** (sum total of genetic information) of many organisms, including humans.

DNA, deoxyribonucleic acid, is a polymer biological molecule (made up of a string of monomers). Each monomer is made up of three major components. Deoxyribose is a sugar. (RNA, ribonucleic acid, is a close relative that uses the sugar ribose.) In addition to the sugar, each DNA nucleotide has a phosphate group and one of four bases named A, C, G, or T. (In RNA, they are A, C, G, and U.)

The bases form hydrogen bonds, A with T and C with G (in RNA, U replaces T), and the monomers are linked with covalent bonds—sugar, phosphate, sugar, phosphate. The space between the outer bonds pulls the structure into a spiral formation as shown in Fig. 3-11.

Fig. 3-11 Structure of DNA.

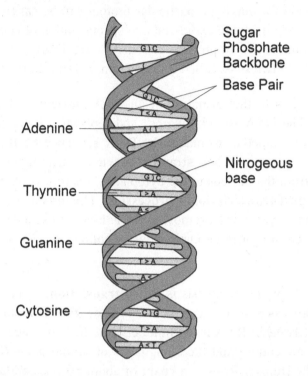

DNA encodes all the information needed for differentiating cells to form an organism and to produce proteins to keep an organism functioning. This information is encoded into genes that exist on the DNA molecules. A **gene** is a length of DNA that encodes a particular protein. Each protein the cell synthesizes performs a specific function in the cell. The function of one protein, or the function of a group of proteins, is called a trait (i.e., hair color, production of an enzyme, specialization of a cell type, etc.).

DNA Replication

In order to replicate, a portion of a DNA molecule unwinds, separating the two halves of the double helix. (This separation is aided by the enzyme helicase.) Another enzyme (DNA polymerase) binds to each strand and moves along them as it collects nucleotides using the original DNA strands as templates. The new strand is complementary to the original template and forms a new double helix with one of the parent strands. If no errors occur during DNA synthesis, the result is two identical double helix molecules of DNA.

DNA carries the information for making all the proteins a cell can make. The DNA information for making a particular protein can be called the gene for that protein. Genetic traits are expressed and specialization of cells occurs, as a result of the combination of proteins encoded by the DNA of a cell. Protein synthesis occurs in two steps called transcription and translation.

Transcription refers to the formation of an RNA molecule, which corresponds to a gene. The DNA strand "unzips" and replicates; individual RNA nucleotides are strung together to match the DNA sequence by the enzyme RNA polymerase. The new RNA strand (known as messenger RNA or **mRNA)** migrates from the nucleus to the cytoplasm, where it is modified in a process known as **post-transcriptional processing**. This processing prepares the mRNA for protein synthesis by removing the non-coding sequences. In the processed RNA, each unit of three nucleotides or **codon** encodes a particular amino acid.

The next phase of protein synthesis is called **translation**. In order for the protein synthesis process to continue, a second type of RNA is required transfer RNA or **tRNA**. Transfer RNA is the link between the "language" of nucleotides (codon and anticodon) and the "language" of amino acids (hence the word "translation"). Transfer RNA is a chain of about 80 nucleotides. At one point along the tRNA chain, there are three unattached bases, which are called the anticodon. This anticodon will line up with a corresponding codon during translation. Each tRNA molecule also has an attached specific amino acid.

Translation occurs at the ribosomes. A ribosome is a structure composed of proteins and ribosomal RNA (rRNA). A ribosome attaches to the mRNA strand at a particular codon known as the start codon. This codon is only recognized by a particular initiator tRNA. The ribosome continues to add tRNA whose anticodons make complementary bonds with the next codon on the mRNA string, forming a peptide bond between amino acids as each amino acid is held in place by a tRNA. At the end of the translation process, a terminating codon stops the synthesis process and the protein is released.

Mutations

Mutations can and do occur in the DNA replication and protein synthesis process all the time. All the DNA of every cell of every organism is copied repeatedly to form new cells for growth, repair, and reproduction. A mutation can result from an error that randomly occurs during replication. Mutations can also result from damage to DNA caused by exposure to certain chemicals, such as some solvents or the chemicals in cigarette smoke, or by radiation, such as ultraviolet radiation in sunlight or x-rays. In many cases, the living organism has mechanisms to recognize and destroy dysfunctional mutated cells, or proteins. Some mutations develop into malformations (i.e., a freckle or a mole) or diseases, such as cancer, and cause damage to the living organism. However, these mutated traits are never passed on to offspring unless the mutation happens within the gametes (sex cells) that are produced in the sex organs (ovaries or testes).

Structural and Regulatory Genes

Genes encode proteins of two varieties. **Structural genes** code proteins that form organs and structural characteristics. **Regulatory genes** code proteins that determine functional or physiological events, such as growth. These proteins regulate when other genes start or stop encoding proteins, which in turn produce specific traits.

Cell Division

The process of cell reproduction is called **cell division**. The process of cell division centers on the replication and separation of strands of **DNA**.

Structure of Chromosomes

Chromosomes are long chains of subunits called **nucleosomes.** Each nucleosome is composed of a short length of DNA wrapped around a core of small proteins called **histones**. The combination of DNA with histones is called **chromatin**. Each nucleosome is about 11 nm in diameter (a nanometer is one billionth of a meter) and contains a central core of eight histones with the DNA double helix wrapped around them. Each gene spans dozens of nucleosomes. The DNA plus histone strings are then tightly packed and coiled, forming chromatin.

Fig. 3-12 A Chromosome.

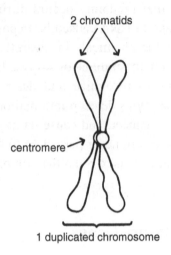

In a cell that is getting ready to divide, each strand of chromatin is duplicated. The two identical strands (called **chromatids**) remain attached to each other at a point called the **centromere**. During cell division, the chromatin strands become more tightly coiled and packed forming a chromosome, which is visible in a light microscope. At this stage, a chromosome consists of two identical chromatids, held together at the centromere, giving each chromosome an **X** shape.

Within the nucleus, each chromosome pairs with another of similar size and shape. These pairs are called **homologs**. Each set of homologous chromosomes has a similar genetic constitution, but the genes are not necessarily identical. Different forms of corresponding genes are called **alleles**.

Fig. 3-13 Paired Homologous Chromosomes.

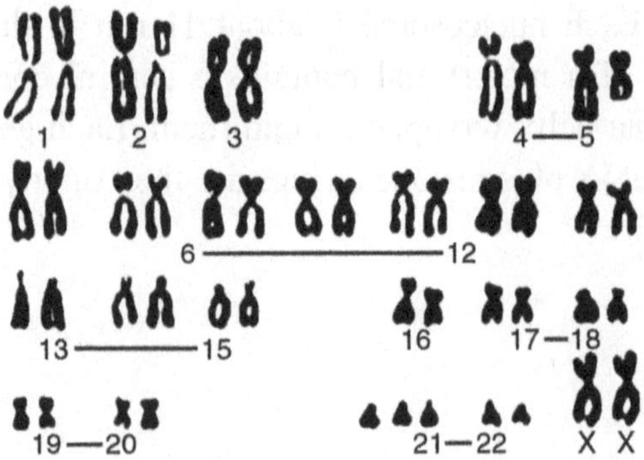

The Cell Cycle

A cell that is going to divide progresses through a particular sequence of events ending in cell division, which produces two daughter cells. This is known as the **cell cycle** (see Fig. 3-14). The time taken to progress through the cell cycle differs with different types of cells, but the sequence is the same. Cells in many tissues never divide.

Fig. 3-14 The Cell Cycle. Interphase includes the G1, S, and G2 phases. The cell division phase includes mitosis and cytokinesis.

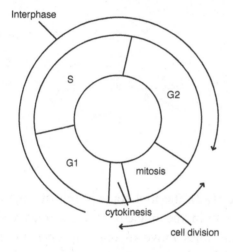

There are two major periods within the cell cycle: interphase and mitosis (also called the M phase or cell division phase). **Interphase** is the period when the cell is active in carrying on its function. Interphase is divided into three phases. During the first phase, the G_1 **phase**, metabolism and protein synthesis are occurring at a high rate, and most of the growth of the cell occurs at this time. The cell organelles are produced (as necessary) and undergo growth during this phase. During the second phase, the **S phase**, the cell begins to prepare for cell division by replicating the DNA and proteins necessary to form a new set of chromosomes. In the final phase, the G_2 **phase**, more proteins are produced, which will be necessary for cell division, and the centrioles (which are integral to the division process) are replicated as well. Cell growth and function occur through all the stages of interphase.

Mitosis

Mitosis is the process by which a cell distributes its duplicated chromosomes so that each daughter cell has a full set of chromosomes. Mitosis progresses through four phases: prophase, metaphase, anaphase, and telophase (see Fig. 3-13).

Fig. 3-15 Mitosis. See explanations of numbered steps below.

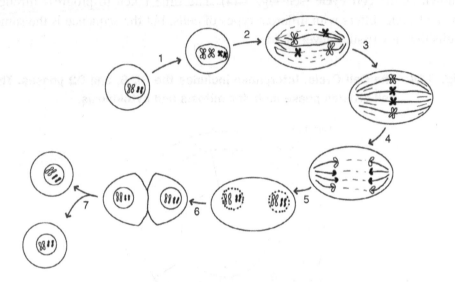

During **prophase (1, 2)**, the first stage of mitosis, the chromatin condenses into chromosomes within the nucleus and becomes visible through a light microscope. The centrioles move to opposite ends of the cell, and **spindle fibers** begin to extend from the centromeres of each chromosome toward the center of the cell. At this point, although the chromosomes become visible, the nucleolus no longer is. During the second part of prophase, the nuclear membrane dissolves and the spindle fibers attach to the centromeres forming a junction called a **kinetochore.** The chromosomes then begin moving in preparation for the next step, metaphase.

During **metaphase (3)**, the spindle fibers pull the chromosomes into alignment along the equatorial plane of the cell, creating the metaphase plate. This arrangement ensures that one copy of each chromosome is distributed to each daughter cell.

During **anaphase (4)**, the chromatids are separated from each other when the centromere divides. Each former chromatid is now called a chromosome. The two identical chromosomes move along the spindle fibers to opposite ends of the cell. **Telophase (5)** occurs as nuclear membranes form around the chromosomes. The chromosomes disperse through the new nucleoplasm, and are no longer visible as chromosomes under a standard microscope. The spindle fibers disappear. After telophase, the process of **cytokinesis (6)** produces two separate cells (7).

Cytokinesis differs somewhat in plants and animals. In animal cells, a ring made up of the protein actin surrounds the center of the cell and contracts. As the actin ring contracts, it pinches the cytoplasm into two separate compartments. Each cell's plasma membrane seals, making two distinct daughter cells. In plant cells, a cell plate forms across the center of the cell and extends out towards the edges of the cell. When this plate reaches the edges, a cell wall forms on either side of the plate, and the original cell then splits into two.

Mitosis, then, produces two nearly identical daughter cells. (Cells may differ in distribution of mitochondria or because of DNA replication errors, for example.) Organisms (such as bacteria) that reproduce asexually, do so through the process of mitosis.

Meiosis

Meiosis is the process of producing four daughter cells, each with single unduplicated chromosomes **(haploid)**. The parent cell is **diploid**, that is, it has a normal set of paired chromosomes. Meiosis goes through a two-stage process resulting in four new cells, rather than two (as in mitosis). Each cell has half the chromosomes of the parent. Meiosis occurs in reproductive organs, and the resultant four haploid cells are called **gametes** (egg and sperm). When two haploid gametes fuse during the process of fertilization, the resultant cell has one chromosome set from each parent, and is diploid. This process allows for the huge genetic diversity available among species.

Two distinct nuclear divisions occur during meiosis, reduction (or meiosis 1, steps **1 to 5** in Fig. 3-16), and division (or meiosis 2, steps **6 to 10**). **Reduction** affects the **ploidy** (referring to haploid or diploid) level, reducing it from 2n to n (i.e., diploid to haploid). **Division** then distributes the remaining set of chromosomes in a mitosis-like process.

Fig. 3-16 Meiosis. See explanations of numbered steps in text.

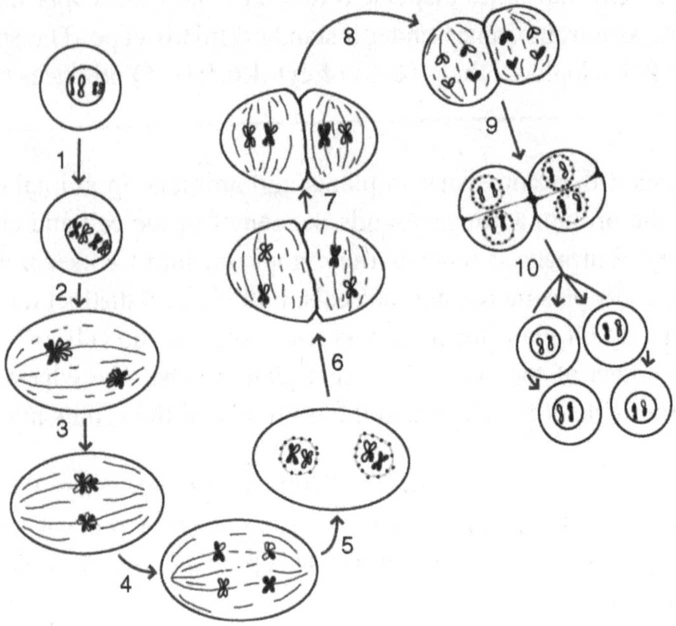

The phases of meiosis 1 are similar to the phases of mitosis, with some notable differences. As in mitosis, chromosome replication (**1**) occurs before prophase; then during prophase 1 (**2**), homologous chromosomes pair up and join at a point called a **synapse** (this happens only in meiosis). The attached chromosomes are now termed a tetrad, a dense four-stranded structure composed of the four chromatids from the original chromosomes. At this point, some portions of the chromatid may break off and reattach to another chromatid in the tetrad. This process, known as **crossing over**, results in an even wider array of final genetic possibilities.

The nuclear membrane disappears during late prophase (or prometaphase). Each chromosome (rather than each chromatid) develops a kinetochore, and as the spindle fibers attach to each chromosome, they begin to move.

In metaphase 1 (**3**), the two chromosomes (a total of four chromatids per pair) align themselves along the equatorial plane of the cell. Each homologous pair of chromosomes contains one chromosome from the mother and one from the father from the original sexual production of that organism. When the homologous pairs orient at the cell's center in preparation for separating, the chromosomes randomly sort. The resulting cells from this meiotic division will have a mixture of chromosomes from each parent. This increases the possibilities for variety among descendent cells.

Anaphase 1 (**4**) occurs next as the chromosomes move to separate ends of the cell. This phase differs from the anaphase of mitosis where one of each chromosome pair (rather than one chromatid) separates. In telophase 1 (**5**), the nuclear envelope may or may not form, depending on the type of organism. In either case, the cell then proceeds to meiosis 2.

The nuclear envelopes dissolve (if they have formed) during prophase 2 (**6**) and spindle fibers form again. All else proceeds as in mitosis, through metaphase 2 (**7**), anaphase 2 (**8**), and telophase 2 (**9**). Again, as in mitosis, each chromosome splits into two chromatids. The process ends with cytokinesis (**10**), forming four distinct gamete cells.

Biosynthesis

Biosynthesis is the process of producing chemical compounds by living things. Organisms need to produce membranes and organelles as well as proteins that regulate cellular activity and molecules that store energy. Biosynthesis takes reactant molecules and with added energy (and often the action of enzymes) produces the products that are needed for cell or organism function.

Protein synthesis, as described earlier, involves DNA, RNA, and the processes of transcription and translation. Protein synthesis is a biosynthetic process vital to all life functions.

Enzymes are protein molecules that act as catalysts for organic reactions. (A catalyst is a substance that lowers the activation energy of a reaction. A catalyst is not consumed in the reaction.) Enzymes do not make reactions possible that would not otherwise occur under the right energy conditions, but they lower the activation energy, which increases the rate of the reaction.

Fig. 3-17 Effect of Enzyme on a Reaction. Adding an enzyme lowers the activation energy for a reaction.

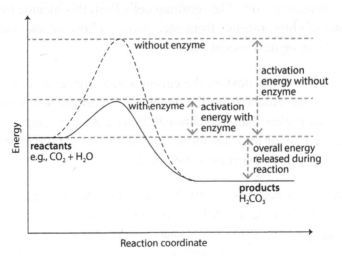

Enzymes are named ending with the letters -*ase*, and usually begin with a syllable describing the catalyzed reaction (i.e., hydrolase catalyzes hydrolysis reactions, lactase catalyzes the breakdown of the sugar lactose). Thousands of reactions occur within cells, each controlled by one or more enzymes. Enzymes are synthesized within the cell at the ribosomes, as all proteins are.

Enzymes are effective catalysts because of their unique shapes. Each enzyme has a uniquely shaped area, called its **active site**. For each enzyme, there is a particular substance known as its **substrate**, which fits within the active site (like a hand in a glove). When the substrate is seated in the active site, the combination of two molecules is called the **enzyme-substrate complex**. An enzyme can bind to two substrates and catalyze the formation of a new chemical bond linking the two substrates. An enzyme may also bind to a single substrate and catalyze the breaking of a chemical bond releasing two products. Once the reaction has taken place, the unchanged enzyme is released.

The operation of enzymes lowers the energy needed to initiate cellular reactions. However, the completion of the reaction may either require or release energy. Reactions requiring energy are called endothermic reactions. Reactions that release energy are called exothermic reactions. Endothermic reactions can take place in a cell by being coupled to the breakdown of ATP or a similar molecule. Exothermic reactions are coupled to the production of ATP or another molecule with high-energy chemical bonds.

Fig. 3-18 Enzyme Reaction. 1: enzyme; 2: substrate; 3 & 4: enzyme-substrate complex; 5: products.

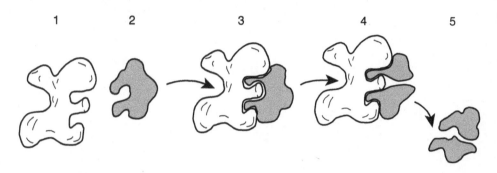

Environmental conditions within the cell, such as high temperature or acidity, may inhibit an enzymatic reaction. These conditions may change the shape of the active site and render the enzyme ineffective.

CHAPTER 4

Structure and Function of Plants and Animals; Genetics

CHAPTER 4

STRUCTURE AND FUNCTION OF PLANTS AND ANIMALS; GENETICS

PLANTS (BOTANY)

Most of us commonly recognize plants as organisms that produce their own food through the process of photosynthesis. (Some bacteria are also photosynthetic.) However, the plant kingdom is divided into several classifications according to physical characteristics.

Vascular plants (tracheophytes) have tissue organized in such a way as to conduct food and water throughout their structure. These plants include some that produce seeds (such as corn or roses) as well as those that do not produce any seeds (such as ferns). **Nonvascular** plants (bryophytes), such as mosses, lack special tissue for conducting water or food. They produce no seeds or flowers and are generally only a few centimeters in height.

Another method of classifying plants is according to their method of reproduction. **Angiosperms** are plants that produce flowers as reproductive organs. **Gymnosperms**, on the other hand, produce seeds without flowers. These include conifers (cone-bearers) and cycads.

Plants that survive only through a single growing season are known as **annuals**. Other plants are **biennial**; their life cycle spans two growing seasons. **Perennial** plants continue to grow year after year.

Plant Anatomy

Plants have structures with attributes that equip them to thrive in their environment. Angiosperms and Gymnosperms differ mostly in the structure of their stems and reproductive organs. Gymnosperms are mostly trees, with woody instead of herbaceous stems. Gymnosperms do not produce flowers; instead they produce seeds in cones or cone-like structures.

Fig. 4-1 A Typical Flowering Plant (angiosperm). See descriptions of numbered structures in text.

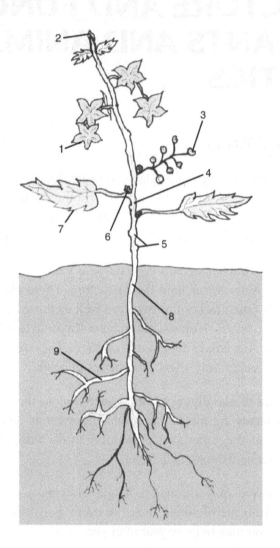

Angiosperms

The shoot system of angiosperms includes the **stem (4)**, **leaves (7)**, **flowers (1)**, and **fruit (3)**, as well as growth structures such as **nodes (5)** and **buds (6)** and the **shoot apex (2)** (see Fig. 4-1). In addition, the **primary root (8)** and **secondary root (9)** provide nourishment of water and minerals and structural support for the plant as it grows. The signature structure of an angiosperm is the **flower (1)** (see Fig. 4-2), the primary reproductive organ. Before the flower blooms, it is enclosed within the **sepals (1-a)**, small, green, leaf-like structures, which fold back to reveal the flower **petals (1-b)**. The petals usually are brightly colored; their main function is to attract insects and birds, which may be necessary to the process of pollination. The short branch of the stem, which supports the flower, is called the **pedicel (1-c)**.

Fig. 4-2 Typical Flower. See numbered explanations in text.

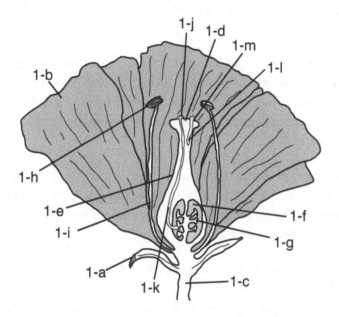

Usually (but depending on the species), a single flower will have both male and female reproductive organs. The **pistil** is the female structure, and includes the stigma, style, ovary, and ovules. The **stigma (1-d)** is a sticky surface at the top of the pistil, which traps pollen grains. The stigma sits above a slender vase-like structure, the **style (1-e)**, which encloses the ovary. The **ovary (1-f)** is the hollow, bulb-shaped structure in the lower interior of the pistil. (After seeds have formed, the ovary will ripen and become fruit.) Within the ovary are the **ovules (1-g)**, small round cases each containing one or more egg cells. (If the egg is fertilized, the ovule will become a seed.) In the process of meiosis in the ovule, an egg cell is produced, along with smaller bodies known as polar nuclei. The polar nuclei will develop into the endosperm of the seed when fertilized by sperm cells.

The male structure is the stamen, consisting of the **anther (1-h)** atop the long, hollow **filament (1-i)**. The anther has four lobes and contains cells (microspore mother cells) that become pollen. Some mature **pollen grains**

(1-j) are conveyed (usually by wind, birds, or insects) to a flower of a compatible species, where they stick to the stigma. The stigma produces chemicals, which stimulate the pollen to burrow into the style, forming a hollow **pollen tube (1k)**. This tube is produced by the tube **nucleus (1-l)**, which has developed from a portion of the pollen grain. The pollen tube extends down toward the ovary. Behind the tube nucleus are two **sperm nuclei (1-m)**. When the sperm nuclei reach the ovule, one will join with an egg cell, fertilizing it to become a zygote (the beginning cell of the embryo). The other sperm nucleus merges with the polar bodies forming the endosperm, which will feed the growing embryo.

Fruit is a mature ovary, which contains the seeds (mature fertilized ovules). The fruit provides protection for the seeds, as well as a method to disburse them. For instance, when ripened fruit is eaten by animals the seeds are discarded or excreted in the animal's waste, transferring the seed to a new location for germination.

Each **seed** contains a tiny embryonic plant, stored food, and a seed coat for protection. When the seed is exposed to proper moisture, temperature, and oxygen, it germinates (begins to sprout and grow into a new plant). Stored food in a seed is found in the cotyledon.

The **stem** is the main support structure of the plant. The stem produces leaves and lateral (parallel with the ground) branches. The stem is also the main organ for transporting food and water to and from the leaves. In some cases the stem also stores food; for instance, a potato is a tuber (stem) that stores starch. The stem also contains meristem tissue.

Most of the stem tissue is made up of **vascular tissue**, including two varieties—**xylem** and **phloem**. Xylem tissue is composed of long tubular cells, which transport water up from the ground to the branches and leaves. Phloem tissue, made of stacked cells connected by sieve plates (which allow nutrients to pass from cell to cell), transports food made in the leaves (by photosynthesis) to the rest of the plant.

The **leaf** is the primary site of photosynthesis in most plants. Most leaves are thin, flat, and joined to a branch or stem by a petiole (a small stem-like extension). The petiole houses vascular tissue, which connects the veins in the leaf with those in the stem.

Fig. 4-3   Cross Section of a Leaf.

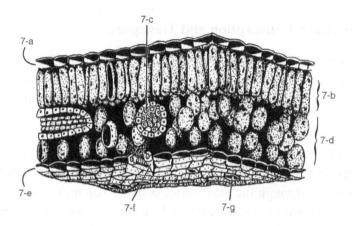

The **cuticle**, which maintains the leaf's moisture balance, covers most leaf surfaces. Considering a cross-section of a leaf (Fig. 4-3), the outermost layer is the **epidermis (7-a, e)**. The epidermis is generally one cell thick. It secretes the waxy cuticle and protects the inner tissue of the leaf.

The mesophyll is composed of several layers of tissue between the upper and lower epidermis. The uppermost, the **palisade layer (7-b)**, contains vertically aligned cells with numerous chloroplasts. The arrangement of these cells maximizes the potential for exposure of the chloroplasts to needed sunlight. Most photosynthesis occurs in this layer.

The sugars produced by photosynthesis are transported throughout the plant via the **vascular bundles (7-c)** of xylem and phloem. The vascular bundles make up the veins in the leaf.

The next layer beneath the palisade cells is the **spongy layer (7-d)**, a layer of parenchyma cells separated by large air spaces. The air spaces allow for the exchange of gases (carbon dioxide and oxygen) for photosynthesis.

On the underside of the leaf there are openings ringed by **guard cells (7-f)**. The openings are called stomata **(7-g)** (or stomates). The stomata serve to allow moisture and gases (carbon dioxide and oxygen) to pass in and out of the leaf, thus facilitating photosynthesis.

Plant Physiology

Water and Mineral Absorption and Transport

Although plants produce their own sugars and starches for food, they must obtain water, carbon dioxide, and minerals from their environment. Vascular plants have well-developed systems for absorption and transport of water and minerals.

Water is essential to all cells of all plants, so plants must have the ability to obtain water and transport water molecules throughout their structure. Most water is absorbed through the plant's root system, then makes its way in one of two pathways toward the xylem cells, which will transport water up the stem and to the leaves and flowers. The first pathway is for water to seep between the epidermal cells of the roots and between the parenchyma cells of the cortex. When water reaches the endodermal tissue, it enters the cells and is pushed through the vascular tissue toward the xylem.

A second pathway is for the water to pass through the cell wall and plasma membrane. Water travels along this intracellular route through channels in the cell membranes (plasmodesmata), until it reaches the xylem.

Once water reaches the xylem, hydrogen bonding between water molecules (known as **cohesion**) causes tension that pulls water through the water column up through the stem and on to the leaves (known as the **cohesion-tension process**). Some water that has traveled up through the plant to the leaves is evaporated in a process known as **transpiration**. As water is evaporated it causes a siphoning effect (like sucking on a straw), which continues to pull water up from the root xylem, through the length of the plant and to the leaves.

Food Translocation and Storage

Food is manufactured by photosynthesis mostly in the leaves. The rest of the plant must have this food (carbohydrates) imported from the leaves. The leaves have source cells, which store the manufactured sugars. The food molecules are transferred from the source cells to phloem tissue through active transport (energy is expended to move molecules across the plasma membrane against the concentration gradient—from low concentration to high concentration). Once in the phloem, the sugars begin to build up, causing osmosis to occur (water enters the phloem lowering the sugar concentration). The entrance of water into the phloem causes pressure, which pushes the water-sugar solution through **sieve plates** that join the cells. This pressure thrusts the water-sugar solution to all areas of the plant, making food available to all cells in the plant.

Plant Reproduction and Development

The reproductive cycle of plants occurs through the alternation of haploid (n) and diploid (2n) phases. Haploid cells have one complete set of chromosomes (n). Diploid cells have two sets of chromosomes (2n). Diploid and haploid stages are both capable of undergoing mitosis in plants. The diploid generation is known as a **sporophyte**. The reproductive organs of the sporophyte produce **gametophytes** through the process of meiosis. Gametophytes may be male or female and are haploid. The male gametophyte produces **sperm** (**male gamete**); the female produces an **egg cell** (**female gamete**). When a sperm cell fertilizes an egg cell (haploid cells join to form a diploid cell) they produce a **zygote**. The zygote will grow into an **embryo**, which resides within the growing seed.

Asexual Plant Reproduction

Some plants may also reproduce through **vegetative propagation**—an asexual process. Asexual reproduction occurs through mitosis only (it does not involve gametes), and produces offspring genetically identical to the parent. While sexual reproduction leads to genetic variation and adaptation, asexual reproduction of a plant with a desirable set of genetic traits preserves these intact in successive generations. Many plants reproduce through a combination of sexual and asexual reproduction, reaping the advantages of each.

ANIMALS (ZOOLOGY)

The animal kingdom includes a wide variety of phyla that have a range of body plans. This range includes certain invertebrates with relatively simple body plans as well as highly complex vertebrates (including humans). There are specific characteristics that differentiate animals from other living things. Organisms in the animal kingdom share the following traits:

1. Animal cells do not have cell walls or plastids.

2. Adult animals are multicellular with specialized tissues and organs.
3. Animals are heterotrophic (they do not produce their own food).
4. Animal species are capable of sexual reproduction, although some are also capable of asexual reproduction (ex. hydra).
5. Animals develop from embryonic stages.

In addition to the above traits, most adult animals have a symmetrical anatomy. Adult animals can have either radial symmetry (constituent parts are arranged radiating symmetrically about a center point) or bilateral symmetry (the body can be divided along a center plane into equal, mirror-image halves). There are a few exceptions to this rule, including the adult sponge whose body is not necessarily symmetrical. While there is wide variation in the physical structure of animals, the animal kingdom is usually divided into two broad categories—invertebrates and vertebrates. There are many more species of invertebrates than vertebrates.

Invertebrates are those species having no internal backbone structure; **vertebrates** have internal backbones. Invertebrates include sponges and worms, which have no skeletal structure at all, and arthropods, mollusks, crustaceans, etc., which have exoskeletons. In fact, there are many more species of invertebrates than vertebrates (about 950,000 species of invertebrates and only about 40,000 species of vertebrates).

Animal Anatomy

Tissues

Like all multicellular organisms, animal bodies contain several kinds of tissues, made up of different cell types. Differentiated cells may organize into specialized tissues performing particular functions. There are eight major types of animal tissue:

1. **Epithelial tissue** consists of thin layers of cells. Epithelial tissue makes up the layers of skin, lines ducts, and the intestine, and covers the inside of the body cavity. Epithelial tissue forms the barrier between the environment and the interior of the body.

2. **Connective tissue** covers internal organs and composes ligaments and tendons. This tissue holds tissues and organs together, stabilizing the body structure.

3. **Muscle tissue** is divided into three types—smooth, skeletal, and cardiac. **Smooth** muscle makes up the walls of internal organs and functions in involuntary movement (breathing, etc.). **Skeletal** muscle attaches bones of the skeleton to each other and surrounding tissues. Skeletal muscle's function is to enable voluntary movement. **Cardiac** muscle is the tissue forming the walls of the heart. Its strength and electrical properties are vital to the heart's ability to pump blood.

4. **Bone tissue** is found in the skeleton and provides support, protection for internal organs, and ability to move as muscles pull against bones.

5. **Cartilage tissue** reduces friction between bones, and supports and connects them. For example, it is found at the ends of bones and in the ears and nose.

6. **Adipose tissue** is found beneath the skin and around organs providing cushioning, insulation, and fat storage.

7. **Nerve tissue** is found in the brain, spinal cord nerves, and ganglion. It carries electrical and chemical impulses to and from organs and limbs to the brain. Nerve tissue in the brain receives these impulses and sustains mental activity.

8. **Blood tissue** consists of several cell types in a fluid called plasma. It flows through the blood vessels and heart, and is essential for carrying oxygen to cells, fighting infection, and carrying nutrients and wastes to and from cells. Blood also has clotting capabilities, which preserve the body's functions in case of injury.

Tissues are organized into organs, and organs function together to form systems, which support the life of an organism. Studying these systems allows us to understand how organisms thrive within their ecosystem.

Systems

Many different body plans exist amongst animals, and each type of body plan includes systems necessary for the organism to live. Our discussion of systems here focuses on those found in most vertebrates. Vertebrates are highly complex organisms with several systems working together to perform the functions necessary to life. These include the digestive, gas exchange, skeletal, nervous, circulatory, excretory, and immune systems.

Digestive System

The **digestive system** (see Fig. 4-4) serves as a processing plant for ingested food. The digestive system in animals generally encompasses the processes of **ingestion** (food intake), **digestion** (breaking down of ingested particles into molecules that can be absorbed by the body), and **egestion** (the elimination of indigestible materials). In most vertebrates, the digestive organs are divided into two categories, the **alimentary canal**, and the **accessory organs**. The alimentary canal is also known as the **gastrointestinal (or GI) tract** and includes the mouth, pharynx, esophagus, stomach, small intestine, large intestine, rectum, and anus. The accessory organs include the teeth, tongue, salivary glands, liver, gallbladder, and pancreas.

Fig. 4-4 Human Digestive System.

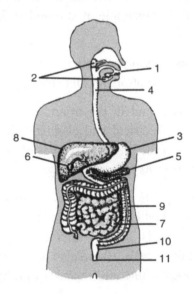

The **mouth** (oral cavity, **1**) is the organ of ingestion and the first organ of digestion in the GI tract. The first step in digestion in many vertebrates occurs as food is chewed. Chewing is the initial step in breaking down food into particles of manageable size. Chewing also increases the surface area of the food and mixes it with saliva, which contains the starch-digesting enzyme amylase. Saliva is secreted by the **salivary glands (2)**. Chewed food is then swallowed and moved toward the **stomach (3)** by peristalsis (muscle contraction) of the **esophagus (4)**. The stomach is a muscular organ that stores incompletely digested food. The stomach continues the mechanical and chemical breakdown of food particles begun by the chewing process. The lining of the stomach

CHAPTER 4: STRUCTURE AND FUNCTION OF PLANTS AND ANIMALS; GENETICS

secretes mucous to protect it from the strong digestive chemicals necessary in the digestive process. The stomach also secretes digestive enzymes and hydrochloric acid which continue the digestive process to the point of producing a watery soup of nutrients, which then proceeds through the pyloric sphincter into the small intestine (the duodenum). The **pancreas (5)** and **gall bladder (6)** release more enzymes into the small intestine, the site where the final steps of digestion and most absorption occurs. The cells lining the **small intestine (7)** have protrusions out into the lumen of the intestine called **villi**. Villi provide a large surface area for absorption of nutrients. Nutrients move into the capillaries through or between the cells making up the villi. The enriched blood travels to the **liver (8)**, where some sugars are removed and stored. The indigestible food proceeds from the small intestine to the **large intestine (9)** where water is absorbed back into the body. The waste (feces) is then passed through the **rectum (10)** and excreted from the **anus (11)**.

Fig. 4-5 Human Respiratory System.

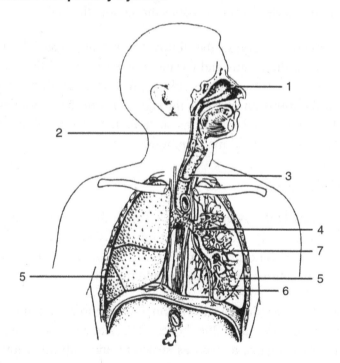

Respiratory or Gas Exchange System

Also known as the **respiratory system** (see Fig. 4-5), the **gas exchange system** is responsible for the intake and processing of gases required by an organism, and for expelling gases produced as waste products. In humans, air is taken in primarily through the nose (although gases may be inhaled through the mouth, the nose is better at filtering out pollutants in the air). The **nasal passages (1)** have a mucous lining to capture foreign particles. This lining is surrounded by epithelial tissue with embedded capillaries, which serve to warm the entering air. Air then passes through the **pharynx (2)** and into the **trachea (3)**. The trachea includes the windpipe or **larynx** in its upper portion, and the **glottis**, an opening allowing gases to pass into the two branches known as the bronchi. The glottis is guarded by a flap of tissue, the **epiglottis**, which prevents food particles from entering the bronchial tubes. The **bronchi (4)** lead to the two **lungs (5)** where they branch out in all directions into smaller tubules known as **bronchioles (6)**. The bronchioles end in **alveoli (7)**, thin-walled air sacs, which are the site of gas exchange. The bronchioles are surrounded by capillaries, which bring blood with a high density of carbon dioxide and a low concentration of oxygen from the pulmonary arteries. At the alveoli, the carbon dioxide diffuses from the blood into the alveoli and oxygen diffuses from the alveoli into the blood. The oxygenated blood is carried away to tissues throughout the body.

All living organisms require the ability to exchange gases, and there are several variations to the means and organs utilized for this life process. Invertebrates such as the earthworm are able to absorb gases through their skin. Insects rely on the diffusion of gases through holes in the exoskeleton known as spiracles. In single-celled organisms such as the amoeba, diffusion of gases occurs directly through the plasma membrane.

Musculoskeletal System

The **musculoskeletal system** provides the body with structure, stability, and the ability to move. By definition, the musculoskeletal system is unique to vertebrates, although some invertebrates (such as mollusks and insects) have external support structures (exoskeletons) and muscle.

In humans, the musculoskeletal system is composed of joints, ligaments, cartilage, muscle groups, and 206 bones. The skeleton provides protection for the soft internal organs, as well as structure and stability allowing for an upright stature and movement. Bones also perform the important function of storing calcium and phosphates, and producing red blood cells within the bone marrow. The 206 bones forming the human skeleton are linked with movable joints, and joined by muscle systems controlling movement.

Skeletal muscles are voluntary—they are activated by command from the nervous system. **Smooth muscle** lines most internal organs, protecting their contents and function, and generally contracting without conscious intent. For instance, the involuntary (automatic) contraction of smooth muscle in the esophagus and lungs facilitates digestion and respiration. **Cardiac muscle** is unique to the heart. It is involuntary muscle (like smooth muscle), but cardiac muscle also has unique features, which cause it to "beat" rhythmically. Cardiac muscle cells have branched endings that interlock with each other, keeping the muscle fibers from ripping apart during their strong contractions. In addition, electrical impulses travel in waves from cell to cell in cardiac muscle, causing the muscle to contract in a coordinated way with a rhythmic pace.

Nervous System

The **nervous system** is a communication network that connects the entire body of an organism, and provides control over bodily functions and actions. Nerve tissue is composed of nerve cells known as **neurons**. Neurons carry impulses via electrochemical responses through their **cell body** and **axon** (long root-like appendage of the cell). Nerve cells exist in networks, with axons of neighbor neurons interacting across small spaces **(synapses)**. Chemical neurotransmitters send messages along the nerve network causing responses specific to varying types of nerve tissue. The nervous system allows the body to sense stimuli and conditions in the environment and respond with necessary reactions. **Sensory organs**—skin, eyes, nose, ears, etc.—transmit signals in response to environmental stimuli to the **brain**, which then conveys messages via nerves to glands and muscles, which produce the necessary response.

Fig. 4-6   A Typical Neuron.

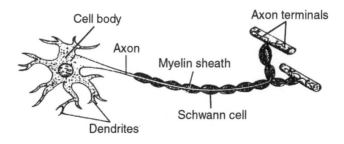

The human nervous system (and that of many mammals) is anatomically divided into two systems—the central nervous system and the peripheral nervous system. The following outline shows the components of each portion of the nervous system:

I. **Central Nervous System** (CNS)—two main components, the **brain** and the **spinal cord**. These organs control all other organs and systems of the body. The spinal cord is a continuation of the brain stem, and acts as a conduit of nerve messages.

II. **Peripheral Nervous System** (PNS)—a network of nerves throughout the body.

  A. **Sensory Division**

   1. **visceral sensory nerves**—carry impulses from body organs to the CNS

   2. **somatic sensory nerves**—carry impulses from body surface to the CNS

  B. **Motor Division**

   1. **somatic motor nerves**—carries impulses to skeletal muscle from the CNS

   2. **autonomic**

     a. **sympathetic** nervous system—carries impulses that stimulate organs

     b. **parasympathetic** nervous system—carries impulses back from organs

Fig. 4-7 Human Nervous System.

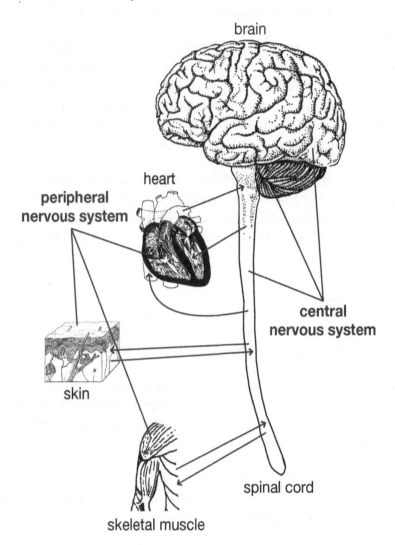

The brain of vertebrates has three major divisions: the forebrain, midbrain, and hindbrain. The **forebrain** is located most anterior, and contains the **olfactory lobes** (sense of smell), **cerebrum** (controls sensory and motor responses, memory, speech, and most factors of intelligence), as well as the **thalamus** (integrates senses), **hypothalamus** (involved in hunger, thirst, blood pressure, body temperature, hostility, pain, pleasure, etc.), and **pituitary gland** (releases various hormones). The **midbrain** is between the forebrain and hindbrain and contains the **optic lobes** (visual center connected to the eyes by the optic nerves). The **hindbrain** consists of the **cerebellum** (controls balance, equilibrium, and muscle coordination) and the **medulla oblongata** (controls involuntary response such as breathing and heartbeat).

Within the brain, nerve tissue is grayish in color and is called **gray matter**. The nerve cells, which exist in the spinal cord and throughout the body, have insulation covering their axons. This insulation (called the **myelin sheath**) speeds electrochemical conduction within the axon of the nerve cell. Since the myelin sheath gives this tissue a white color, it is called **white matter**. The myelin sheath is made up of individual cells called Schwann cells.

The nervous systems of vertebrates and some invertebrates are highly sophisticated, providing conscious response and unconscious controls. However, the nervous systems of some species of invertebrates (such as jellyfish) are relatively simple networks of neurons that control only some aspects of their body functions.

Circulatory System

The **circulatory system** is the conduit for delivering nutrients and gases to all cells and for removing waste products from them.

In invertebrates, the circulatory system may consist entirely of diffusion in the gastrovascular cavity, or it may be an **open circulatory system** (where blood directly bathes the internal organs), or a **closed circulatory system** (where blood is confined to vessels).

Closed circulatory systems are also typical of vertebrates. In vertebrates, **blood** flows throughout the circulatory system within **vessels.** Vessels include **arteries, veins,** and **capillaries**. The pumping action of the **heart** (a hollow, muscular organ) forces blood in one direction throughout the system. In large animals, valves within the heart, and some of the vessels in limbs, keep blood from flowing backwards (being pulled downward by gravity).

Blood carries many products to cells throughout the body, including minerals, infection-fighting white blood cells, nutrients, proteins, hormones, and metabolites. Blood also carries dissolved gases (particularly oxygen) to cells and waste gases (mainly carbon dioxide) away from cells. The process of cellular metabolism is a fundamental process of life and cannot proceed without a continuous supply of oxygen to every living cell within the body.

Fig. 4-8 Human Circulatory System. Blood flows from the heart through arteries to the capillaries throughout the body and returns via the veins.

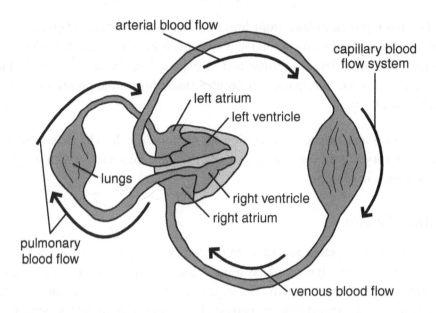

Capillaries (tiny vessels) surround all tissues of the body and exchange carbon dioxide for oxygen. Oxygen is carried by **hemoglobin** (containing iron) in red blood cells. Oxygen enters the blood in the lungs and travels to the heart, then through **arteries** (larger vessels that carry blood away from the heart), then **arterioles** (small arteries), to capillaries. The blood picks up carbon dioxide waste from the cells and carries it through capillaries, then **venules** (small veins) and **veins** (vessels that carry blood toward the heart), back to the heart and on to the lungs. Thus, blood is continually cycled.

Excretory System

The **excretory system** is responsible for collecting waste materials and transporting them to organs that expel them from the body. There are many types of waste that must be expelled from the body, and there are many organs involved in this process.

The primary excretory organs of most vertebrates are the kidneys. The **kidneys** filter metabolic wastes from the blood and excrete them as **urine** into the urinary tract. The urinary tract carries the fluid that is eventually expelled from the body. Urine is typically 95% water, and may contain urea (formed from breakdown of proteins), uric acid (formed from breaking down nucleic acids), creatinine (a byproduct of muscle contraction), and various minerals and hormones.

The **liver** produces **bile** from broken-down pigments and chemicals (often from pollutants and medications) and secretes it into the small intestine, where it proceeds to the large intestine and is expelled in the feces. The liver also breaks down some nitrogenous molecules (including some proteins), excreting them as urea.

The **lungs** are the sites of excretion for carbon dioxide. The **skin** is an accessory excretory organ; salts, urea, and other wastes are secreted with water from sweat glands in the skin.

Immune System

The **immune system** functions to defend the body from infection by bacteria and viruses. The **lymphatic system** is the principal infection-fighting component of the immune system. The organs of the lymphatic system in humans and other higher invertebrates include the lymph, lymph nodes, spleen, thymus, and tonsils. **Lymph** is a collection of excess fluid that is absorbed from between cells into a special system of vessels, which circulates through the lymphatic system and finally dumps into the bloodstream. Lymph also collects plasma proteins that have leaked into interstitial fluids.

Lymph nodes are small masses of lymph tissue whose function is to filter lymph and produce lymphocytes. **Lymphocytes** and other cells are involved in the immune system. Lymphocytes begin in bone marrow as stem cells and are collected and distributed via the lymph nodes. There are two classes of lymphocytes: B cells, and T cells. **B cells** emerge from the bone marrow mature, and produce **antibodies**, which enter the bloodstream. These antibodies find and attach themselves to foreign **antigens** (i.e., toxins, bacteria, foreign cells, etc.). The attachment of an antibody to an antigen marks the pair for destruction.

The **spleen** contains some lymphatic tissue, and is located in the abdomen. It filters larger volumes of lymph than nodes can handle. The **tonsils** are a group of lymph cells connected together and located in the throat.

The **thymus** is another mass of lymph tissue, which is active only through the teen years, fighting infection and producing T cells. **T cells** mature in the thymus gland. Some T cells (like B cells) patrol the blood for antigens, but T cells are also equipped to destroy antigens themselves. T cells also regulate the body's immune responses.

Homeostatic Mechanisms

All living cells, tissues, organs, and organisms must maintain a tight range of physical and chemical conditions in order for them to live. Conditions such as temperature, pH, water balance, sugar levels, and so on, must be monitored and controlled in order to keep them within the accepted ranges that will not inhibit life. When the conditions of an organism are within acceptable ranges, it is said to be in homeostasis. Organisms have a special set of mechanisms that serve to keep them in **homeostasis**. Homeostasis is a state of dynamic equilibrium, which balances forces tending toward change with forces acceptable for life functions.

Homeostasis is achieved mostly by actions of the sympathetic and parasympathetic nervous systems by a process known as **feedback control**. For instance, when the body undergoes physical activity, muscle action causes a rise in temperature. If not checked, rising temperature could destroy cells. In this instance, the nervous system detects rising temperature and reacts with a response that causes sweat glands to produce sweat. The evaporation of sweat cools the body.

There are many instances of feedback control. These take effect when any situation arises that may drive levels outside the normal acceptable range. In other words, the homeostatic mechanism is a reaction to a stimulus. This reaction, called a **feedback response**, is the production of some counterforce that levels the system.

Hormonal Control in Homeostasis and Reproduction

Hormones are chemicals produced in the endocrine glands of an organism, which travel through the circulatory system and are taken up by specific targeted organs or tissues, where they modify metabolic activities.

Hormonal control occurs through one of two processes. The first is the **mobile receptor mechanism**. A hormone is manufactured in response to a particular **stimulus**. The hormone (for instance, a **steroid**) enters the bloodstream from one of the ductless endocrine glands that manufacture hormones.

The steroid passes through the cell membrane of the targeted cell and enters the cytoplasm. The hormone combines with a particular protein known as a receptor, creating the **hormone-receptor complex**. This complex enters the nucleus and binds to a DNA molecule causing a gene to be transcribed. The mRNA molecule leaves the nucleus for the endoplasmic reticulum, where it encodes a particular protein. The protein migrates to the site of the stimulus and counteracts the source of the stimulus. The result is homeostasis, a balance of the counterproductive forces.

The second process targets receptors on a cell's membrane. A particular **receptor** exists on the membrane when the cell is in a particular condition (for instance, containing an excess of glucose). When the hormone binds with the receptor on the membrane, the receptor changes its form. This triggers a chain of events within the cytoplasm resulting in the production or destruction of proteins, thus moderating the conditions.

Hormones control many physiological functions, from digestion, to conscious responses and thinking, to reproduction. In humans, for instance, women of childbearing age have a continuous cycle of hormones. The hormone cycle causes the release of eggs at specific times. If the egg is fertilized a different combination of hormones stimulates a chain of events that promotes the development of the embryo.

Animal Reproduction and Development

Reproduction in multicellular animals is a complex process that generally proceeds through the steps of **gametogenesis** (gamete formation) and then **fertilization**.

Gametes are the sex cells formed in the reproductive organs—sperm and eggs. When a sperm of one individual combines with the egg cell of another, the resulting cell is known as a **zygote**. A **zygote** then develops into a new individual. In the case of **spermatogenesis** (sperm formation), diploid **primary spermatocytes** are formed from special cells **(spermatogonia)** in the testes. The primary spermatocytes then undergo meiosis I, forming haploid **secondary spermatocytes** with a single chromosome set. (Please see the section on meiosis in Chapter 3.) The secondary spermatocytes go through meiosis II, forming **spermatids**, which are haploid. These spermatids then develop into the **sperm cells**.

In human female reproductive organs, egg cells are formed through a similar process known as **oogenesis**. **Primary oocytes** are typically present in great number in the female's ovaries at birth. Primary oocytes undergo meiosis I, forming one **secondary oocyte** and one smaller **polar body**. Both the secondary oocyte and the polar body undergo meiosis II; the polar body producing two polar bodies (not functional cells), and the oocyte producing one more polar body and one haploid **egg cell**. The egg cell is now ready for fertilization, and if there are sperm cells present, the egg may be fertilized forming a diploid cell with a new combination of chromosomes, the zygote.

PRINCIPLES OF HEREDITY (GENETICS)

The process by which characteristics pass from one generation to another is known as **inheritance**. The study of the principles of heredity (now called genetics) advanced greatly through the experimental work of **Gregor Mendel** (c. 1865). Mendel studied the relationships between traits expressed in parents and offspring, and the hereditary factors that caused expression of traits.

Mendel systematically bred pea plants to determine how certain hereditary traits passed from generation to generation. First, he established true-breeding plants, which produce offspring with the same traits as the parents. For example, the seeds of pea plants with yellow seeds would grow into plants that produced yellow seeds. Green seeds grow into plants that produce green seeds. Mendel named this first generation of true-breeding plants the parent or P_1 **generation**; he then bred the plant with yellow seeds and the plant with green seeds. Mendel called the first generation of offspring the F_1 **generation**. The F_1 generation of Mendel's yellow seed/green seed crosses contained only yellow seed offspring.

Mendel continued his experiment by crossing two individuals of the F_1 generation to produce an F_2 **generation**. In this generation, he found that some of the plants (one out of four) produced green seeds. Mendel performed hundreds of such crosses, studying some 10,000 pea plants, and was able to establish the rules of inheritance from them. The following are Mendel's main discoveries:

- Parents transmit hereditary factors (now called **genes**) to offspring. Genes then produce a characteristic, such as seed-coat color.
- Each individual carries two copies of a gene, and the copies may differ.
- The two genes an individual carries act independently, and the effect of

one may mask the effect of the other. Mendel coined the terms geneticists still use: "dominant" and "recessive."

Modern Genetics

We now know that **chromosomes** carry all the genetic information in most organisms. Most organisms have corresponding pairs of chromosomes that carry genes for the same traits. These pairs are known as **homologous chromosomes**. Genes that produce a given trait exist at the same position (or **locus**) on homologous chromosomes. Each gene may have different forms, known as **alleles**. For instance, yellow seeds and green seeds arise from different alleles of the same gene. A gene can have two or more alleles, which differ in their nucleotide sequence. That difference can translate into proteins that function differently, resulting in variations of the trait.

Sexual reproduction (meiosis) produces gamete cells with one-half the genetic information of the parents (paired chromosomes are separated and sorted independently). Therefore, each gamete may receive one of any number of combinations of each parent's chromosomes.

In addition, a trait may arise from one or more genes. (However, because one-gene traits are easiest to understand, we will use them for most of our examples.) If a trait is produced from a gene or genes with varying alleles, several possibilities for traits exist. The combination of alleles that make a particular trait is the **genotype**, while the trait expressed is the **phenotype**.

An allele is considered **dominant** if it masks the effect of its partner allele. The allele that does not produce its trait when present with a dominant allele is **recessive**. That is, when a dominant allele pairs with a recessive allele, the expressed trait is that of the dominant allele.

A **Punnett square** is a notation that allows us to easily predict the results of a genetic cross. In a Punnett square, a letter is assigned to each gene. Uppercase letters represent dominant traits, while lowercase letters represent recessive traits (a convention begun by Mendel). The possible alleles from each parent are noted across the top and side of a box diagram; then the possible offspring are represented within the internal boxes. If we assign the allele that produces yellow seeds the letter **Y**, and the allele that produces green seeds y, we can represent Mendel's first cross between pea plants (**YY** × **yy**) by the following Punnett square:

	Y	Y
y	Yy	Yy
y	Yy	yy

One parent pea plant had green seeds (green seeds is its phenotype), so it must not have had any of the dominant genes for yellow seeds (**Y**); therefore it must have the genotype **yy**. If the second parent had one allele for yellow and one for green, then some of the offspring would have inherited two genes for green. Since Mendel started with true-breeding plants, we may deduce that one parent had two genes for green seeds (**yy**) and the other two genes for yellow seeds (**YY**).

When both alleles for a given gene are the same in an individual (such as **YY** or **yy**), that individual is **homozygous** for that trait. Furthermore, the individual's genotype can be called homozygous. Both of the above parents (P_1) were homozygous. The children in the F_1 generation all have one dominant gene (**Y**) and one recessive gene (**y**), their phenotype is yellow, and their genotype is **Yy**. When the two alleles for a given gene are different in an individual (**Yy**), that individual is said to be **heterozygous** for that trait; its genotype is heterozygous.

Breeding two F_1 offspring from the example above produces the following Punnett square of a double heterozygous (both parents **Yy**) cross:

	Y	Y
Y	YY	Yy
y	Yy	yy

Through this Punnett square, we can determine that three-fourths of the offspring will produce yellow seeds. This is consistent with Mendel's findings. However, there are two different genotypes represented among the yellow seed offspring. One-half of the offspring were heterozygous yellow (**Yy**), while one-fourth were homozygous yellow (**YY**).

The previous example shows a **monohybrid cross**—a cross between two individuals where only one trait is considered. Mendel also experimented with crossing two parents while considering two separate traits, a **dihybrid cross**.

The laws investigated by Mendel form the basis of modern genetics. However, Mendel's laws now incorporate modern terminology (i.e., "genes" rather than "hereditary factors," etc.).

The Law of Segregation

The first law of Mendelian genetics is the **law of segregation**. The law of segregation states that traits are expressed from a pair of genes in the individual (on homologous chromosomes). Each parent provides one chromosome of every pair of homologous chromosomes. Paired chromosomes (and thus corresponding genes) separate and randomly recombine during gamete formation.

The Law of Dominance

Mendel determined that one gene was usually dominant over the other. This is the **law of dominance**, Mendel's second law of inheritance. In Mendel's experiments, the first generation produced no plants with green seeds, leading him to recognize the existence of genetic dominance. The yellow-seed allele was clearly dominant.

The Law of Independent Assortment

Mendel also investigated whether genes for one trait always were linked to genes for another. In other words, Mendel experimented not only with pea seed-coat color, but also with pea-plant height (and a number of other traits in peas and other plants). He wanted to determine whether if the parent plant had green seeds and was tall, all plants with green seeds would be tall. These dihybrid cross experiments demonstrated that most traits were independent of one another. That is, a pea plant could be green and tall or green and short, yellow and tall, or yellow and short. In most cases, genes for traits randomly sort into pairs (although some genes lie close to others on a chromosome and can therefore be inherited together). Since homologous chromosomes separate and independently sort in gamete formation, alleles are also separated and independently sorted, an assertion known as the **law of independent assortment**.

The following Punnett square demonstrates independent assortment. **Y** stands for the allele for yellow color, **y** for the allele for green, **T** for the allele tall, and **t** for short:

	TY	Ty	tY	ty
TY	TTYY	TTYy	TtYY	TtYy
Ty	TTYy	TTyy	TtYy	Ttyy
tY	TtYY	TtYy	ttYY	ttYy
ty	TtYy	Ttyy	ttYy	ttyy

Incomplete Dominance

Some traits have no genes that are dominant and instead produce offspring that are a mix of the two parents. For instance, in snapdragons a plant with red flowers crossed with a plant with white flowers produces offspring with pink flowers. This is known as **incomplete dominance**. Neither white nor red is dominant over the other. In incomplete dominance, the conventional way to symbolize the alleles is with a capital letter designating the trait (in this case C for color) and a superscript designating the allele choices (in this case ^{R} for red, ^{W} for white), making the possible alleles C^R and C^W. The following Punnett square represents the incomplete dominance of the allele for red flowers (C^R), the allele for white as (C^W), and the combination resulting in pink as ($C^R C^W$).

	C^R	C^R
C^W	$C^R C^W$	$C^R C^W$
C^W	$C^R C^W$	$C^R C^W$

In this case, two plants, one with white flowers, one with red, cross to form all pink flowers. If two of the heterozygous offspring of this cross are then bred, the outcome of this cross ($C^R C^W \times C^R C^W$) will be:

	C^R	C^W
C^R	$C^R C^R$	$C^R C^W$
C^W	$C^R C^W$	$C^W C^W$

One-fourth of the offspring will be red one-half pink, and one-fourth white, a 1:2:1 ratio.

Multiple Alleles

In the instances above, two possible alleles exist in a species, so the genotype will be a combination of those two alleles. There are some instances where more than two choices of alleles are present. For instance, for human blood types there is a dominant allele for type A blood, another dominant allele for type B blood as well as a recessive allele for neither A nor B, known as O blood. There are three different alleles and they may combine in any way. In multiple-allele crosses, it is conventional to denote the chromosome by a letter (in this case **I** for dominant, **i** for recessive), with a subscript letter representing the allele types (in this case **A, B,** or **O**). The alleles for A and B blood are co-dominant, while the allele for O blood is recessive. The possible genotypes and phenotypes, then, are as follows:

genotype	phenotype
$I^A I^A$	Type A blood
$I^B I^B$	Type B blood
$I^B i^O$	Type B blood
$I^A i^O$	Type A blood
$I^A I^B$	Type AB blood
$i^O i^O$	Type O blood

[Note: There is another gene responsible for the Rh factor that adds the + or − to the blood type.]

Linkage

While Mendel had established the law of independent assortment, later study of genetics by other scientists found that this law was not always true. In studying fruit flies, for instance, it was found that some traits are always inherited together; they were not independently sorted. Traits that are inherited together are said to be **linked**. Genes are portions of chromosomes, so most traits produced by genes on the same chromosome are inherited together. (The chromosomes are independently sorted not the individual genes.)

However, an exception to this rule complicates the issue. During metaphase of meiosis I, when homologous chromosomes line up along the center of the dividing cell, some pieces of the chromosomes break off and move from one chromosome to another (change places). This random breaking and reforming of homologous chromosomes allows genes to change the chromosome they are linked to, thus changing the genome of that chromosome. This process, known as **crossing over**, adds even more possibility of variation of traits among species. It is more likely for crossing over to occur between genes that do not lie close together on a chromosome than between those that lie close together.

Gender is determined in an organism by a particular homologous pair of chromosomes. The symbols **X** and **Y** denote the sex chromosomes. In mammals and many insects, the male has an **X** and **Y** chromosome **(XY)**, while the female has two **X**'s **(XX)**. Genes that are located on the gender chromosome Y will only be seen in males. It would be considered a **sex-limited trait**. An example of a sex-limited trait is bar coloring in chickens that occurs only in males.

Some traits are **sex-linked**. In sex-linked traits, more males **(XY)** develop the trait because males have only one copy of the **X** chromosome. Females have a second **X** gene, which may carry a gene coding for a functional protein for the trait in question that may counteract a recessive trait. These traits (for example, hemophilia and colorblindness) occur much more often in males than females.

Still other traits may be **sex-influenced**. In this case, the trait is known as autosomal—it only requires one recessive gene to be expressed if there is no counteracting dominant gene. A male with one recessive allele will develop the trait, whereas a female would require two recessive genes to develop it. An example of a sex-influenced trait is male-pattern baldness.

Polygenic Inheritance

While the best-studied genetic traits arise from alleles of a single gene, most traits, such as height and skin color, are produced from the expression of more than one set of genes. Traits produced from interaction of multiple sets of genes are known as **polygenic traits**. Polygenic traits are difficult to map and difficult to predict because of the varied effects of the different genes for a specific trait.

CHAPTER 5
Ecology and Population Biology

CHAPTER 5

ECOLOGY AND POPULATION BIOLOGY

ECOLOGY

Ecology is the study of how organisms interact, and how they influence or are influenced by their physical **environment**. The word "ecology" is derived from the Greek term *oikos* (meaning "home" or "place to live") and *ology* (meaning "the study of"), so ecology is a study of organisms in their home. This study has revealed a number of patterns and principles that help us understand how organisms relate to their environment. First, however, it is important to grasp some basic vocabulary used in ecology.

The study of ecology centers on the ecosystem. An **ecosystem** is a group of populations found within a given locality, plus the inanimate environment around those populations. A **population** is the total number of a single species of organism found in a given ecosystem. Typically, there are many populations of different species within a particular ecosystem. The term **organism** refers to an individual of a particular species. Each species is a distinct group of individuals that are able to interbreed (mate), producing viable offspring. Although species are defined by their ability to reproduce, they are usually described by their morphology (their anatomical features).

Populations that interact with each other in a particular ecosystem are collectively termed a **community**. For instance, a temperate forest community includes pine trees, oaks, shrubs, lichen, mosses, ferns, squirrels, deer, insects, owls, bacteria, fungi, etc.

The part of the Earth that includes all living things is called the **biosphere**. The biosphere also includes the **atmosphere** (air), the **lithosphere** (ground), and the **hydrosphere** (water).

A **habitat** refers to the physical place where a species lives. A species' habitat must include all the factors that will support its life and reproduction. These factors may be **biotic** (i.e., living – food source, predators, etc.) and **abiotic** (i.e., nonliving – weather, temperature, soil features, etc.).

A species' **niche** is the role it plays within the ecosystem. It includes its physical requirements (such as light and water) and its biological activities (how it reproduces, how it acquires food, etc.). One important aspect of a species' niche is its place in the food chain.

Ecological Cycles

Every species within an ecosystem requires resources and energy in varying forms. The interaction of organisms and the environment can be described as cycles of energy and resources that allow the community to flourish. Although each ecosystem has its own energy and nutrient cycles, these cycles also interact with each other to form bioregional and planetary biological cycles.

The **energy cycle** supports life throughout the environment. There are also several **biogeochemical cycles** (the water cycle, the carbon cycle, the nitrogen cycle, the phosphorous cycle, the rock cycle, etc.), which are also important to the health of ecosystems. A biogeochemical cycle is the system whereby the substances needed for life are recycled and transported throughout the environment.

Carbon, hydrogen, oxygen, phosphorous, and nitrogen are called macronutrients; they are used in large quantities by living things. Micronutrients—those elements utilized in trace quantities in organisms—include iodine, iron, zinc, and copper.

Energy Cycle (Food Chain)

Since all life requires the input of energy, the **energy cycles** within the ecosystem are central to its well-being. On Earth, the Sun provides the energy that is the basis of life in most ecosystems. (An exception is the hydrothermal vent communities that derive their energy from the heat of the Earth's core.) Without the constant influx of solar energy into our planetary ecosystem, most life would cease to exist. Energy generally flows through the entire ecosystem in one direction—from producers to consumers and on to decomposers (consumers may also consume decomposers) through the **food chain**.

Photosynthetic organisms—such as plants, some protists, and some bacteria—are the first link in most food chains; they use the energy of sunlight to combine carbon dioxide and water into sugars, releasing oxygen gas (O_2). Photosynthetic organisms are called producers, since they synthesize sugar and starch molecules using the Sun's energy to link the carbons in carbon dioxide. Primary consumers (also known as herbivores) are species that eat photosynthetic organisms. Consumers utilize sugars and starches stored in cells or tissues for energy. Secondary consumers feed on primary consumers, and on the chain goes, through tertiary, quaternary (etc.) consumers. Finally, decomposers (bacteria, fungi, some animals) are species that recycle the organic material found in dead plants and animals back into the food chain.

Fig. 5-1 A Food Chain Pyramid.

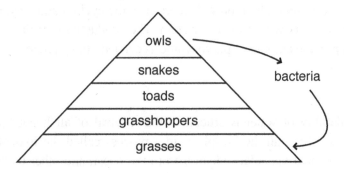

Animals that feed only on other animals are called **carnivores** (meat-eaters), whereas those that consume both photosynthetic organisms and other animals are known as **omnivores**.

The energy cycle of the food chain is subject to the laws of thermodynamics. Energy can neither be created nor destroyed. However, every use of energy is less than 100% efficient; about 10% is lost as heat. When we call photosynthetic organisms producers, we mean that they produce food using the Sun's energy to form chemical bonds in sugars and other biomolecules. Other organisms can use the energy stored in the bonds of these biomolecules.

The steps in the food chain are also known as **trophic levels**. Consider the pyramid diagram (Fig. 5-1) as one example of a food chain with many trophic levels. Grasses are on the bottom of the pyramid; they are the producers, the first trophic level. Producers are also known as **autotrophs**, as they produce their own food. Each trophic level is greater in **biomass** (total mass of organisms) than the level above it.

Grasshoppers represent the **second trophic level**, or primary consumers in this example of a food chain. Grasshoppers consume plants and are consumed (in this example) by toads, the secondary consumers, which represent the **third trophic level**. Snakes consume toads, and are in turn consumed by owls— making these the **fourth and fifth trophic levels**. In this example, bacteria are the decomposers that recycle some of the nutrients from dead owls (and other levels) to be reused by the first trophic level.

The pyramid illustrates a food chain; however, in nature it is never actually as simple as shown. Owls consume snakes, but they may also consume toads (a lower level in the pyramid) and fish (from an entirely different pyramid). Thus, within every ecosystem there may be numerous food chains interacting in varying ways to form a **food web**. Furthermore, all organisms produce waste products that feed decomposers. The food web represents the cycling and recycling of both energy and nutrients within the ecosystem. The productivity of the entire web is dependent upon the amount of photosynthesis carried out by photosynthesizers.

Water Cycle

The availability of water is crucial to the survival of all living things. Water vapor circulates through the biosphere in a process called the **hydrologic cycle**. Water is evaporated via solar radiation from the ocean and other bodies of water into clouds. Water is also released into the atmosphere from vegetation (leaves) by transpiration. Some water is also evaporated directly from soil, but most water in the ground flows into underground aquifers, which eventually empty into the oceans. Water above ground flows into waterways, which also eventually flow into the ocean (a process known as runoff). Water vapor is then redistributed over land (and back into oceans as well) via clouds, which release water as precipitation.

The water cycle also has a profound effect on Earth's climate. Clouds reflect the Sun's radiation away from the Earth, causing cool weather. Water vapor in the air also acts as a **greenhouse gas,** reflecting radiation from the Earth's surface back toward the Earth, and therefore trapping heat. The water cycle also intersects nearly all the other cycles of elements and nutrients.

Nitrogen Cycle

Nitrogen is another substance essential to life processes, since it is a key component of amino acids (components of proteins) and nucleic acids. The nitrogen cycle recycles nitrogen. Nitrogen is the most plentiful gas in the atmosphere, making up 78% of the air. However, neither photosynthetic organisms nor animals are able to use nitrogen gas (N_2), which does not readily react with other compounds, directly from the air. Instead a process known as nitrogen fixing makes nitrogen available for absorption by the roots of plants. **Nitrogen fixing** is the process of combining nitrogen with either hydrogen or oxygen, mostly by **nitrogen-fixing bacteria**, or to a small degree by volcanoes and **lightning**.

These nitrogen-fixing bacteria fill a unique and vital niche, by living in the soil and performing the task of combining gaseous nitrogen from the atmosphere with hydrogen, forming ammonium (NH_4^+ ions). (Some cyanobacteria, also called blue-green bacteria, are also active in this process.) Ammonium ions are then absorbed and used by plants. Other types of nitrogen-fixing bacteria live in symbiosis on the nodules of the roots of legumes (beans, peas, clover, etc.), supplying the roots with a direct source of ammonia.

Some plants are unable to use ammonia, instead they use **nitrates**. Some bacteria perform **nitrification**, a process, which further breaks down ammonia into nitrites (NO_2^-), and yet again another type of bacteria converts nitrites into nitrates (NO_3^-).

Nitrogen compounds (such as ammonia and nitrates) are also produced by natural, physical processes such as volcanic activity. Another source of usable nitrogen is lightning, which reacts with atmospheric nitrogen to form nitrates.

In addition, nitrogen passes along through the food chain, and is recycled through decomposition processes. When plants are consumed, the amino acids are recombined and used in a process that passes the nitrogen-containing molecules on through the food chain or web. Animal waste products, such as urine, release nitrogen compounds (primarily ammonia) back into the environment, yet another source of nitrogen. Finally, large amounts of nitrogen are returned to the Earth by bacteria and fungi, which decompose dead plant and animal matter into ammonia (and other substances), a process known as **ammonification**.

Various species of bacteria and fungi are also responsible for breaking down excess nitrates, a process known as **denitrification**, which releases nitrogen gas back into the air. The nitrogen cycle involves cycling nitrogen through both living and non-living entities.

Carbon Cycle

The carbon cycle is the route by which carbon is obtained, used, and recycled by living things. Carbon is an important element contained in the cells of all species. The study of organic chemistry is the study of carbon-based molecules.

Earth's atmosphere contains large amounts of carbon in the form of carbon dioxide (CO_2). Photosynthetic organisms require the intake of carbon dioxide for the process of photosynthesis, which is the foundation of the food chain. Most of the carbon within organisms is derived from the production of carbohydrates through photosynthesis. The process of photosynthesis also releases oxygen molecules (O_2), which are necessary to animal respiration. Animal respiration releases carbon dioxide back into the atmosphere in large quantities.

Since plant cells consist of molecules containing carbon, animals that consume photosynthetic organisms are consuming and using carbon from the photosynthetic organisms. Carbon is passed along the food chain as these animals are then consumed. When animals and photosynthetic organisms die, decomposers, including the detritus feeders, bacteria, and fungi, break down the organic matter. Detritus feeders include worms, mites, insects, and crustaceans, which feed on dead organic matter, returning carbon to the cycle through chemical breakdown and respiration.

Carbon dioxide (CO_2) is also dissolved directly into the oceans, where it is combined with calcium to form calcium carbonate, which is used by mollusks to form their shells. When mollusks die, the shells break down and often form limestone. Limestone is then dissolved by water over time and some carbon may be released back into the atmosphere as CO_2, or used by new ocean species.

Finally, organic matter that is left to decay, may, under conditions of heat and pressure, be transformed into coal, oil, or natural gas (the **fossil fuels**). When fossil fuels are burned for energy, the combustion process releases carbon dioxide back into the atmosphere, where it is available to plants for photosynthesis.

Phosphorus Cycle

Phosphorus is another mineral required by living things. Unlike carbon and nitrogen, which cycle through the atmosphere in gaseous form, phosphorus is only found in solid form, within rocks and soil. Phosphorus is a key component in ATP, NADP (a molecule that, like ATP, stores energy in its chemical bonds), and many other molecular compounds essential to life.

Phosphorus is found within rocks and is released by the process of erosion. Water dissolves phosphorus from rocks, and carries it into rivers and streams. Here phosphorus and oxygen react to form phosphates, which end up in bodies of water. Phosphates are absorbed by photosynthetic organisms in and near the water and are used in the synthesis of organic molecules. As in the carbon and nitrogen cycles, phosphorus is then passed up the food chain and returned through animal wastes and organic decay.

New phosphorous enters the cycle as undersea sedimentary rocks. These rocks are thrust up during the shifting of the Earth's tectonic plates. New rock containing phosphorus is then exposed to erosion and enters the cycling process.

POPULATION GROWTH AND REGULATION

The population growth of a species is regulated by limiting factors that exist within the species' environment. Population growth maintains equilibrium in all species under normal conditions because of these limiting factors. A population's overall growth rate is affected by the birth rate (**natality**) and death rate (**mortality**) of the population. The rate of increase within a population is represented by the birth rate minus the death rate. When the birth rate within a population equals the death rate, the population remains at a constant level.

There are two models of population growth, the exponential curve (or J-curve) and the logistic curve (or the S-curve). The exponential curve represents populations in which there is no environmental or social limit on population size, so the rate of growth accelerates over time. Exponential population growth exists only during the initial population growth in a particular ecosystem, since as the population increases the limiting factors become more influential. In other words, a fish population introduced into a pond would experience exponential population growth until food and space supplies began to limit the population.

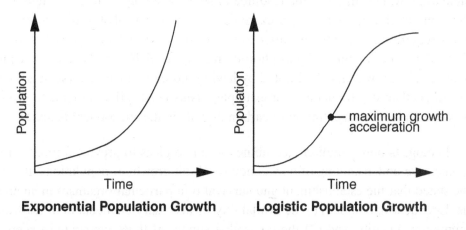

Exponential Population Growth   **Logistic Population Growth**

The logistic curve reflects the effects of limiting factors on population size, where growth accelerates to a point, then slows down. The logistic curve shows population growth over a longer period of time, and represents population growth under normal conditions.

Population growth is directly related to the life characteristics of the population such as the age at which an individual begins to reproduce, the age of death, the rate of growth, etc. For instance, species that grow quickly, mature sexually at an early age, and live a long life would have a population growth rate that exceeds that of species with a short life span and short reproductive span.

In general, most populations have an incredible ability to increase in numbers. A population without limiting factors could overpopulate the world in several generations. It is the limits existing within each ecosystem that keep this from happening.

Limiting Factors

Many factors affect the life of an ecosystem, which may be permanent or temporary. Populations within an ecosystem will be affected by changes in the environment from **abiotic factors** (physical, non-living factors such as fire, pollution, sunlight, soil, light, precipitation, availability of oxygen, water conditions, and temperature) and **biotic factors** (biological factors, including availability of food competition, predator-prey relationships, symbiosis, and overpopulation).

These biotic and abiotic factors are known as **limiting factors** since they will determine how much a particular population within a community will be able to grow. For instance, the resource in shortest supply in an ecosystem may limit population growth. As an example, we know that photosynthetic organisms require phosphorous in order to thrive, so the population growth will be limited by the amount of phosphorous readily available in the environment. Conversely, growth may be limited by having more of an element (such as heat or water) than it can tolerate. For example, plants need carbon dioxide to grow; however, a large concentration of carbon dioxide in the atmosphere is toxic.

Ecologists now commonly combine these two ideas to provide a more comprehensive understanding of how conditions limit growth of populations. It may be stated that the establishment and survival of a particular organism in an area is dependent upon both (1) the availability of necessary elements in at least the minimum quantity, and (2) the controlled supply of those elements to keep it within the limits of tolerance.

Limiting factors interact with each other and generally produce a situation within the ecosystem that supports homeostasis (a steady-state condition). **Homeostasis** is a dynamic balance achieved within an ecosystem functioning at its optimum level. Homeostasis is the tendency of the ecological community to stay the same. However, the balance of the ecosystem can be disturbed by the removal, or decrease, of a single factor or by the addition, or increase, of a factor.

Populations are rarely governed by the effect of a single limiting factor; instead many factors interact to control population size. Changes in limiting factors have a domino effect in an ecosystem, as the change in population size of one species will change the dynamics of the entire community. The number of individuals of a particular species living in a particular area is called the population **density** (number of organisms per area).

Both abiotic and biotic limiting factors exist in a single community; however, one may be dominant over the other. Abiotic limiting factors are also known as **density-independent factors**. That is, they are independent of population density. For instance, the populations around Mount St. Helens were greatly affected by its eruption in 1980, and this effect had nothing to do with the population levels of the area before the eruption. In this situation, the density-independent physical factors dominated the population changes that took place.

Pollution is a major density-independent factor in the health of ecosystems. Pollution is usually a byproduct of human endeavors and affects the air or water quality of an ecosystem with secondary effects. In addition to producing pollution, humans may deliberately utilize chemicals such as pesticides or herbicides to limit growth of particular species. Such chemicals can damage the homeostatic mechanisms within a community, causing a long-term upset in the balance of an ecosystem.

In other situations, biotic factors, called **density-dependent factors** may be the dominant influence on population in a given area. Density-dependent factors include population growth issues and interactions between species within a community.

Within a given area, there is a maximum level the population may reach at which it will continue to thrive. This is known as the **carrying capacity** of the environment. When an organism has reached the carrying capacity of the ecosystem, the population growth rate will level off and show no net growth. Populations also occupy a particular geographic area with suitable conditions. This total area occupied by a species is known as the **range**. Typically, populations will have the greatest density in the center of their range, and lower density at the edges. The area outside the range is known as the area of intolerance for that species, since it is not able to survive there. Environmental changes will affect the size and location of the range, making it a dynamic characteristic.

Over time, species may move in or out of a particular area, a process known as **dispersion**. Dispersion occurs in one of three ways—through **emigration** (permanent one way movement out of the original range), **immigration** (permanent one way movement into a new range), and **migration** (temporary movement out of one range into another, and back). Migration is an important process to many species and communities, since it allows animals that might not survive year round in a particular ecosystem to temporarily relocate for a portion of the year. Therefore, migration gives the opportunity for greater diversity of species in an ecosystem.

Two or more species living within the same area and that overlap niches (their function in the food chain) are said to be in **competition** if the resource they both require is in limited supply. If the niche overlap is minimal (other sources of food are available), then both species may survive. In some cases, one of the species may be wiped out in an area due to competition, a situation called **competitive exclusion**. This is a rare but plausible occurrence.

A **predator** is simply an organism that eats another. The organism that is eaten is known as the **prey**. The **predator/prey** relationship is one of the most important features of an ecosystem. As seen in our study of the energy cycle, energy is passed from lower trophic levels to higher trophic levels, as one animal is consumed by another. This relationship not only provides transfer of energy up the food chain, it also is a population control factor for the prey species. In situations where natural predators are removed from a region, the overpopulation occurring amongst the prey species can cause problems in the population and community. For instance, the hunting and trapping of wolves in the United States has led to an overpopulation of deer (the prey of wolves), which in turn has caused a shortage of food for deer in some areas, causing these deer populations to starve.

When two species interact with each other within the same range, it is known as **symbiosis**. **Amensalism** is one type of symbiosis where one species is neither helped nor harmed while it inhibits the growth of another species. **Mutualism** is another form of symbiosis where both species benefit. **Parasitism** is symbiosis in which one species benefits, but the other is harmed. (Parasites are not predators, since the parasitic action takes a long period of time and may not actually kill the host.)

When the entire population of a particular species is eliminated, it is known as **extinction**. Extinction may be a local phenomenon, the elimination of a population of one species from one area. However, species extinction is a worldwide phenomenon, where all members of all populations of a species die.

The extinction of a single species may also cause a chain reaction of secondary extinctions if other species depend on the extinct species. Conversely, the introduction of a new species into an area can also have a profound effect on other populations within that area. This new species may compete for the niche of native population or upset a predator/prey balance. For example, the brown tree snake (native to Australia) was introduced into islands in the Pacific years ago. (They probably migrated on ships.) The brown snake has caused the extinction of several species of birds on those Pacific islands. The bird populations could not withstand the introduction of this new predator.

Ultimately, the survival of a particular population is dependent on maintaining a **minimal viable population** size. When a population is significantly diminished in size, it becomes highly susceptible to breeding problems and environmental changes that may result in extinction.

Community Structure

Community structure refers to the characteristics of a specified community including the types of species that are dominant, major climatic trends of the region, and whether the community is open or closed. A **closed community** is one whose populations occupy essentially the same range with very similar distributions of density. These types of communities have sharp boundaries called **ecotones** (such as a pond aquatic ecosystem that ends at the shore). An **open community** has indefinite boundaries, and its populations have varying ranges and densities (such as a forest). In an open community, the species are more widely distributed and animals may actually travel in and out of the area.

An open community is often more able to respond to calamity and may be therefore more resilient. Since the boundaries are subtler, the populations of a forest, for instance, may be able to move as necessary to avoid a fire. If, however, a closed community is affected by a traumatic event (for example, a pond being polluted over a short period of time) it may be completely wiped out.

Communities do grow and change over time. Some communities are able to maintain their basic structure with only minor variations for very long periods of time. Others are much more dynamic, changing significantly over time from one type of ecosystem to another. When one community completely replaces another over time in a given area, it is called **succession**. Succession occurs both in terrestrial and aquatic biomes.

Succession may occur because of small changes over time in climate or conditions, the immigration of a new species, disease, or other slow-acting factors. It may also occur in direct response to cataclysmic events such as fire, flood, or human intervention (for example, clearing a forest for farmland). The first populations that move back into a disturbed ecosystem tend to be hardy species that can survive in bleak conditions. These are known as **pioneer communities**.

An example of terrestrial succession occurs when a fire wipes out a forest community. The first new colonization will come from quick growing species such as grasses, which will produce over time a grassland ecosystem. The decay of grasses will enrich the soil, providing fertile ground for germination of seeds for shrubs brought in by wind or animals. The shrubs will further prepare the soil for germination of larger species of trees, which over time will take over the shrub-land and produce a forest community once again.

When succession ends in a stable community, the community is known as the **climax community**. The climax community is the one best suited to the climate and soil conditions, and one that achieves a homeostasis. Generally, the climax community will remain in an area until a catastrophic event (fire, flood, etc.) destroys it.

PART II
Physical Science

CHAPTER 6

Atomic Chemistry

CHAPTER 6

ATOMIC CHEMISTRY

STRUCTURE OF THE ATOM

The study of matter is known as **chemistry**. All matter is made up of **atoms**. The properties of matter are a result of the structure of atoms and their interaction with each other. An **element** is a substance that cannot be broken down into any other substances. The simplest unit of an element that retains the element's characteristics is known as an **atom**. Each atom of a given element has a nucleus containing a unique number of **protons** and usually a similar number of **neutrons**. The nucleus is surrounded by **electrons**.

Fig. 6-1 Atom Illustrations.

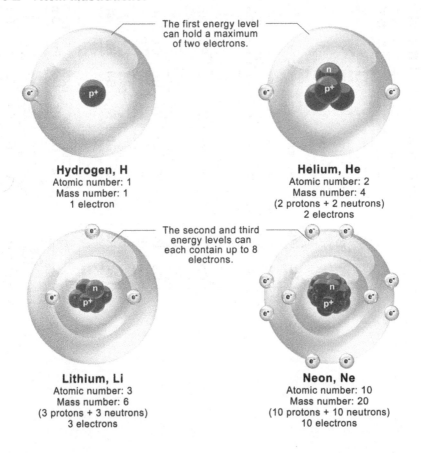

Fig. 6-2 The Periodic Table of the Elements. Elements are listed by atomic number.

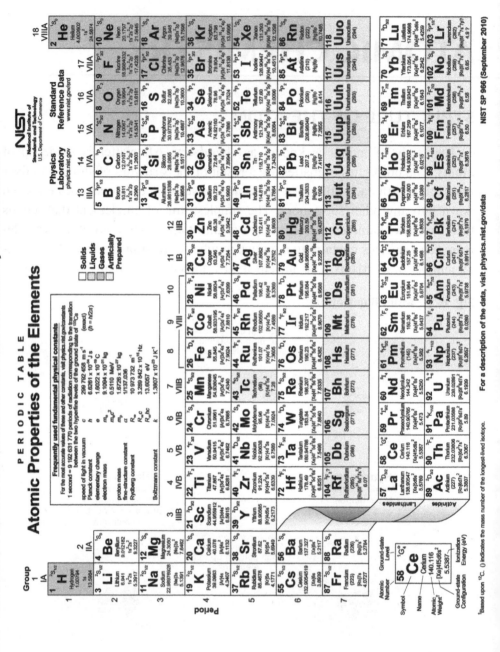

Elements are listed by atomic number on the **Periodic Table of the Elements**. The **atomic number** is the number of protons found in the nucleus of an atom of that element. In an uncharged atom, the number of protons is equal to the number of electrons.

The mass of an atom consists of the cumulative mass of all the particles in the atom, which includes protons, neutrons, and electrons. The mass of the electrons is insignificant relative to the mass of protons or neutrons. Therefore, the **atomic mass** is calculated by adding up the masses of the protons and neutrons. For example, a helium atom consists of two protons and two electrons. In addition, most helium atoms also have 2 neutrons giving it an atomic mass of 4 amu (atomic mass units—the mass of one proton or neutron).

ATOMIC PARTICLE CHART		
NAME	CHARGE	MASS
Proton	+	1 AMU
Electron	−	~0 AMU
Neutron	0	1 AMU

Atoms with the same number of protons but different numbers of neutrons are called **isotopes** of one another. For example, carbon-12 and carbon-14 are the same element (carbon), which is defined as having 6 protons. The difference in the **mass numbers** indicates that carbon-12 has 6 neutrons, while carbon-14 has 8 neutrons. In nature, elements naturally occur as a combination of more than one isotope. The average mass number takes into account the relative frequencies of the different isotopes. The average mass number is also called the **atomic weight**. This number is also the **molar mass** of the element, or the mass in grams of one mole of atoms.

Electrons have a charge of −1, while protons have a charge of +1. Neutrons have no charge. The number of protons in the nucleus of an atom carries a positive charge equal to this number; that is, if an atom's nucleus contains 4 protons, the charge is +4. Since positive and negative charges attract, the positive charges of the nucleus attract an equal number of negatively charged electrons.

Electrons travel freely in a three-dimensional space that may be called an **electron cloud.** Current models of the atom follow the principles of **quantum mechanics,** which predict the probabilities of an electron being in a certain area at a certain time. The orbital represents a probability of finding an electron at a particular location.

Each electron cloud has a particular amount of energy related to it, and is therefore also referred to as an **energy level**. Each energy level has a limited capacity for holding electrons and each energy level requires a different number of electrons to fill it. Lower energy levels (closer to the nucleus) have less capacity for electrons than those farther from the nucleus.

Since electrons are attracted to the nucleus, electrons fill the electron shells closest to the nucleus (lowest energy levels) first. Once a given level is full, electrons start filling the next level out. The outermost occupied energy level of an element is called the **valence shell.** The number of electrons in the valence shell will determine the combinations that this atom will be likely to make with other atoms. Atoms are more stable when every electron is paired and are most stable when their valence shell is full. The tendency for an atom toward stability means that elements having unpaired or partially-filled valence shells will easily gain or lose electrons in order to obtain the most stable configuration.

Electrons give off energy in the form of **electromagnetic radiation** when they move from a higher level, or an excited state, to a lower level. The energy represented by light, using Planck's equation, represents the difference between the two energy levels of the electron.

NUCLEAR REACTIONS AND EQUATIONS

An element is radioactive when the nuclei of its atoms are unstable and spontaneously release one of a variety of subatomic particles in order to form nuclei with higher stability.

Alpha decay occurs when the nucleus of an atom emits a package of two protons and two neutrons, called an **alpha particle** (α), which is equivalent to the nucleus of a helium atom. This usually occurs with elements that have a mass number greater than 60. Alpha decay causes the atom's atomic mass to decrease by four units and the atomic number by two units. For example:

$$^{238}_{92}U \longrightarrow \,^{234}_{90}Th + ^{4}_{2}He$$

(Mass # / #Protons; Alpha Particle)

Beta decay occurs in two forms, positive and negative. A **Beta particle** (β) is a high speed, high energy electron or positron (same particle as an electron but with a reversed – negative – charge). This usually occurs with elements that have a mass number greater than their atomic weight. Beta decay causes the mass number to remain the same but either increases or decreases the atomic number by one. Beta decay converts a neutron into a proton and releases a Beta particle. For example:

Carbon emits a β⁻ gains a proton and becomes Nitrogen.

$$^{14}_{6}C \rightarrow \,^{14}_{7}N + e^{-} + \bar{v}_e$$

(electron; electron neutrino)

Magnesium emits a β⁻ loses a proton and becomes Sodium.

$$^{23}_{12}Mg \rightarrow \,^{23}_{11}Na + e^{+} + v_e$$

(positron; positron neutrino)

Gamma radiation consists of *gamma rays* (γ), which are high-frequency, high-energy electromagnetic radiation that are usually given off in combination with alpha and beta decay. Gamma decay can occur when a nucleus undergoes a transformation from a higher-energy state to a lower-energy state. The resulting atom may or may not be radioactive. Gamma rays are photons, which have neither mass nor charge.

RATE OF DECAY; HALF-LIFE

Half-life is the time it takes for 50 percent of an isotope to decay. Nuclear decay represents a "first-order" reaction in that it depends on the amount of material and the rate constant. For example:

Strontium-85 has a half-life of 65.2 days. How long will it take for 20 grams of strontium-85 to decay into five grams of strontium-85?

Solution:

It takes two half-lives to decrease the amount of strontium-85 from 20 grams to 5 grams.

20 grams decays to 10g in 65.2 days. It takes another 65.2 days for half of that 10g to decay to 5 grams.

Elements such as Uranium are constantly decaying according to their half lives. The quantity of these elements is replenished by the elements with higher atomic numbers decaying down into them.

CHAPTER 7
Chemistry of Reactions

CHAPTER 7

CHEMISTRY OF REACTIONS

COMMON ELEMENTS

Some elements on the Periodic Table of the Elements are more commonly found on Earth than others, and some are more likely to interact with other elements. Properties of each element, such as mass, electronegativity, valence electrons, etc., make a particular element fit for interaction with other elements in a variety of ways. The most common elements encountered in chemical reactions are found on the following table:

Table 7-1. Most Common Elements Found in Chemical Reactions.

Atomic Number	Symbol	Common Name
1	H	Hydrogen
2	He	Helium
6	C	Carbon
7	N	Nitrogen
8	O	Oxygen
11	Na	Sodium
12	Mg	Magnesium
14	Si	Silicon
15	P	Phosphorus
16	S	Sulphur
17	Cl	Chlorine
19	K	Potassium
20	Ca	Calcium
24	Cr	Chromium
26	Fe	Iron
29	Cu	Copper
30	Zn	Zinc
47	Ag	Silver
53	I	Iodine
79	Au	Gold
80	Hg	Mercury
82	Pb	Lead
88	Ra	Radium

CHEMICAL BONDS

Valence properties of atoms provide opportunities for them to bond with other atoms. As discussed in Chapter 6, "Atoms are more stable when every electron is paired and are most stable when the valence shell is full. The tendency for an atom toward stability means that elements having unpaired or partially-filled valence shells will easily gain or lose electrons in order to obtain the most stable configuration." Therefore, certain atoms are easily available to make certain types of bonds under the right conditions with other atoms.

A **covalent bond** between atoms is formed when atoms share electrons. For instance, hydrogen has only one electron, which is unpaired leaving the first valence shell lacking one electron. Oxygen has 6 electrons in the valence shell; it needs 2 more electrons in the valence shell for that shell to be full. It is easy for 2 hydrogen atoms to share their electrons with the oxygen, making the effective valence shells of the oxygen and each hydrogen atom full. The result of the bonding of these three atoms is one molecule of water (H_2O).

A **molecule** is two or more atoms held together by shared electrons (covalent bonds). For example, two hydrogen atoms can join together covalently to form a H_2 molecule. In fact, hydrogen in the air is in the form of H_2. A **compound** is formed when two or more different elemental atoms bond together chemically to form a unique substance (i.e., H_2O, CH_4). These compounds are also molecules. Therefore, all compounds are molecules, but not all molecules are compounds. A covalent bond is the strongest bond due to the sharing of electrons.

Charged atoms are called **ions**. An atom that loses one or more electrons becomes a positively charged particle, or a positive ion. We call this positively charged ion a **cation**. An atom that gains one or more electrons becomes a negative ion, or an **anion**. Positive and negative ions are attracted to each other in an **ionic bond**. An ionic bond is a weak bond, and in fact is considered more of an attraction. Na^+Cl^- (sodium chloride or table salt) is an example of a substance held together by ionic bonds. Na^+Cl^- is considered an ionic substance rather than a molecule since it will quickly dissociate in water.

Some molecules (covalently bonded) have a weak, partial negative charge at one region of the molecule and a partial positive charge in another region. Molecules that have regions of partial charge are called **polar molecules**. For instance, water molecules (which have a net charge of 0) have a partial negative charge near the oxygen atom and a partial positive charge near each of the hydrogen atoms. Thus, when water molecules are close together, their positive regions are attracted to the negatively charged regions of nearby molecules; the negative regions are attracted to the positively charged regions of nearby molecules. The force of attraction between water molecules is called a **hydrogen bond**. A hydrogen bond is a weak chemical bond between molecules.

Fig. 7-2 Hydrogen Bond.

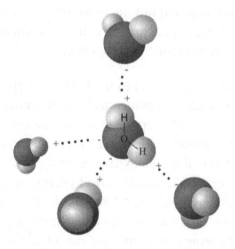

CHEMICAL REACTIONS

Chemical reactions occur when molecules interact with each other to form one or more molecules of another type. There are several categories of chemical reactions. Chemical reactions are symbolized by an equation where the reacting molecules (**reactants**) are shown on one side and the newly formed molecules (**products**) on the other, with an arrow between indicating the direction of the reaction. Some chemical reactions are simple, such as the breakdown of a compound into its components (a **decomposition** reaction):

$$AB \rightarrow A + B$$

A simple **combination** reaction is the reverse of decomposition:

$$A + B \rightarrow AB$$

When one compound breaks apart and forms a new compound with a free reactant, it is called a **replacement** reaction:

$$AB + C \rightarrow AC + B$$

When two compounds break apart and exchange components it is called a double replacement reaction:

$$AB + CD \rightarrow AC + BD$$

Chemical reactions may require an input of energy or they may release energy. Reactions that require energy are called **endothermic** reactions. Reactions that release energy are termed **exothermic**.

All chemical reactions are subject to the **laws of thermodynamics**. The first law of thermodynamics (also known as the law of conservation of matter and energy) states that matter and energy can neither be created nor destroyed. In other words, the sum of matter and energy of the reactants must equal that of the products. The second law of thermodynamics, or the law of increasing disorder (or **entropy**), asserts that all reactions spread energy, which tends to diminish its availability. So, although we know from the first law that the energy must be equal on both sides of a reaction equation, reaction processes also tend to degrade the potential energy into a form that cannot perform any useful work.

CHAPTER 8

Physics

CHAPTER 8

PHYSICS

HEAT

Heat is energy that flows from an object that is warm to an object that is cooler. It is important to understand the difference between heat and temperature. **Temperature** is the measure of the **average kinetic energy** of a substance. The atoms and molecules of all substances are constantly in motion. The energy of the motion of the atoms and molecules in a substance is called its kinetic energy. Temperature is a measure of that energy. The faster the particles in a substance move (more energy), the higher the temperature will be. The slower the particles move (less energy), the lower the temperature.

The theoretical temperature at which particle motion stops is called **absolute zero** (or 0 Kelvin). This temperature has never been reached by any known substance.

When substances come in contact, the hotter (greater energy) substance transfers kinetic energy to the cooler (lower energy) one and heat is expended. Heat is measured in calories or joules.

Energy (including heat) may be transferred from one object to another by three processes—radiation, conduction, and convection. **Radiation** is the transfer of energy via waves. Radiation can occur through matter, or without any matter present. Radiation from the Sun passes through space (very low density of matter) to reach Earth. **Convection** involves the movement of energy by the movement of matter, usually through currents. For instance, convection moves warm air up, while cool air sinks. (Cool air is more dense.) In a fluid the heat will move with the fluid. **Conduction** is movement of energy by transfer from particle to particle. Conduction can only occur when objects are touching. A pan on a stove heats water by conduction. An oven cooks by convection—warming food by movement of hot air. A microwave cooks by radiation—waves travel into food, adding kinetic energy.

Specific Heat

Different substances have different capacities for storing energy. For example, it may take twenty minutes to heat water to 75°C. The same mass of aluminum might require five minutes and the same amount of copper may take only two minutes to reach the same temperature. However, water will retain the kinetic energy (stay hot) longer than the aluminum or copper. The measure of a substance's ability to retain energy is called **specific heat**. Specific heat is measured as the amount of heat needed to raise the temperature of one gram of a substance by one degree Celsius.

THE LAWS OF THERMODYNAMICS

The Laws of Thermodynamics explain the interaction between heat and work (energy) in the universe. The First Law says that matter and energy can neither be created nor destroyed. This law is also known as the **Law of Conservation of Matter and Energy**. The Second Law is known as the **Law of Entropy** and states that whenever energy is exchanged some energy becomes unavailable for use (entropy increases). The Third Law or the **Law of Absolute Zero** says that absolute zero cannot be attained in any system (that is, energy of motion of particles cannot be stopped). The **Zeroth Law of Thermodynamics** is so named because it was accepted by scientists after the first three laws were named but it underlies them all—thus the name "zeroth." The Zeroth Law says that when two bodies are in contact that they will move toward a state of thermodynamic equilibrium—where both bodies eventually reach the same temperature.

STATES OF MATTER

All matter has physical properties that can be observed. These properties affect the way substances react with each other under various conditions. Physical properties include color, odor, taste, strength, hardness, density, and state.

Solids, Liquids, Gases, Plasma

Matter exists in one of four fundamental states—solid, liquid, gas, or plasma. Under most conditions elements will be in the solid, liquid, or gas phase. Plasma only exists in the case of extreme heat and ionization—in this state ions and electrons move about freely, giving plasma properties different from the other three states. The center of the Sun and stars are in the plasma state.

A solid has molecules in fixed positions giving the substance definite shape. Solids have a definite volume. In a solid the molecules are packed and bonded together. The strength of the bonds determines the strength of the solid and its melting point. When heat energy is applied to the bonds, they break apart, the molecules can move about and the substance becomes a liquid.

A liquid has definite volume but not definite shape since the molecules are loosely attracted. The loose attractions between molecules allow for the shape of the substance to mold to its surroundings. When a liquid is cooled, the molecules become bonded and the substance becomes a solid. Heat energy is lost since the molecules are moving less. When heat energy is added to a liquid, the weak attractions holding the molecules together break apart, causing the molecules to move about randomly and the liquid to become a gas.

A gas is a substance with relatively (relative to solids and liquids) large distances and little attraction between molecules. The molecules are free to move about randomly. Gases have no definite shape or volume since temperature and pressure can impact the density.

Different substances have different conditions under which they exist in a solid liquid or gas. The state that a substance exists in at room temperature depends on how the molecules are bonded together.

Fig. 8-1 Solids, liquids, and gases differ in the arrangement and density of molecules.

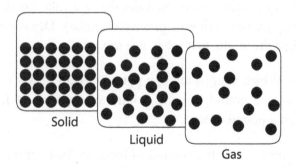

The following are some terms that are important to know related to substances and their states:

- **melting point**—temperature at which a substance changes from solid to liquid form
- **heat of fusion**—heat required to melt 1 kg of a solid at its melting point (also known as enthalpy of fusion)
- **freezing point**—temperature at which a substance changes from liquid to solid
- **boiling point**—temperature at which a substance changes from liquid to gas
- **heat of vaporization**—amount of energy required to change 1 kg of liquid of a substance to a gas (also known as enthalpy of vaporization)
- **evaporation**—escape of individual particles of a substance into gaseous form
- **condensation**—change of a gaseous substance to liquid form
- **diffusion**—mixing of particles in a gas or liquid

Density

Density is the measure of how much matter exists in a given volume. The density of a substance is determined by measuring the mass of a substance and dividing it by the volume $\left(D = \frac{m}{v}\right)$. Substances that are less dense will float when placed in a denser substance. Denser substances will sink in a less dense liquid. It is important to understand that density is a function of mass (amount of matter), not weight (attraction of gravity on mass). Density has to do with how tightly packed the atoms in a substance are.

Consider this: Which weighs more, a pound of lead or a pound of feathers? Neither. A pound is a pound; but the volume of a pound of feathers will be much greater than a pound of lead.

Pressure is a measure of the amount of force applied per unit of area. Pressure exerted on a gas may affect its density, by compressing its volume. Liquids and solids are much less responsive to pressure. **Pascal's principle** states that the pressure exerted on any point of a confined fluid is transmitted unchanged throughout the fluid. Therefore, if you exert pressure on a liquid it will exert that same pressure on its surroundings. **Archimedes' principle** states that when an object is placed in a fluid the object will have a buoyant force equal to the weight of the displaced fluid. Archimedes' principle results in **buoyancy**, a decrease in the measured apparent weight of an object in a fluid due to the net upward force caused by the displaced fluid. This apparent decrease in weight of the object is a result of the force of the displaced fluid acting in an upward direction opposing the object.

Gravity

The **mass** of an object refers to the amount of matter that is contained by the object; however, the **weight** of an object is the force of gravity acting upon that object. Mass is related to how much material is there while weight refers to the pull of the Earth (or any other planet or object) upon that material.

The mass of an object (measured in kg) will be the same no matter where in the universe the object is located. Mass is never altered by location, the pull of gravity, speed or even the existence of other forces. A 51 kg object will have a mass of 51 kg whether it is located on Earth, the Moon, or anywhere else. The amount of mass and the gravitational field of the Earth (or the Moon, etc.) imparts weight to an object. The weight of an object, however, will vary according to where in the universe the object is. Weight depends upon the mass of the celestial body that is exerting the gravitational field and the distance from the center of the gravitational field of the body upon the mass experiencing the field.

Gravity is a property of all matter; all matter exerts a gravitational force on all other matter. Gravity acts at a distance and attracts bodies of matter toward each other. The gravity of the Moon affects the water in the Earth's oceans, causing the tides. In the study of atomic particles, there is even a weak force of gravity between all particles.

The gravity on the Earth is greater than the gravity on the Moon, since the Earth has much more matter or mass than the Moon. The force of gravity on an object caused by the mass of the Earth equals the mass of the object (m) times the acceleration caused by gravity (g). The equation is:

$$F = mg$$

This acceleration caused by gravity on Earth, g (more commonly called the acceleration of gravity) equals 9.8 m/s² in the metric system and 32 ft/s² in the English system.

The weight of an object is the measurement of the force of gravity on that object. When you weigh something on a scale, the weight is actually the force of gravity on that object:

$$Weight = mg$$

The mass of the Moon is less than the mass of the Earth, so the acceleration of gravity (g) is less on the Moon than the Earth. If you put an object on the Moon and weighed it, its weight would be $\frac{1}{6}$ the weight on Earth. In other words, a 180-pound man would only weigh 30 pounds on the Moon.

The fundamental units of measure used in the metric system are the meter, kilogram, and second or mks system. To get the weight of an object in the metric system, you multiply the mass in kilograms by the acceleration of gravity (9.8 m/s²), resulting in the units of kg • m/s² or Newtons.

The universal gravitational law, first described by Sir Isaac Newton, states that *the force of gravity between two objects is proportional to the product of the masses of the objects and inversely proportional to the square of the distance between them*—in simple English—as objects get further apart, the effect of gravity drops dramatically. The equation below shows this relationship . . . the mass of one body is designated as M, the mass of a second as m, and the distance between them is r; the force of attraction between the two bodies is F.

$$F = G\frac{Mm}{r^2}$$

where G is the universal gravitational constant $G = 6.67 \times 10^{-11} \, N\left(\frac{m^2}{kg^2}\right)$ (Newton-meter squared per kilogram squared).

If you drop an object relatively near the Earth, it will speed up according to the acceleration of gravity (g). When you let go of the object, its velocity is zero. Since $g = 32$ ft/s² $= 9.8$ m/s², the velocity will be 32 ft/s² (9.8 m/s²) after one second. Because the object is accelerating, the velocity after 2 seconds will be 2 s × 32 ft/s/s = 64 ft/s² (19.6 m/s²). After 10 seconds, the velocity will be 10 s × 32 ft/s/s = 320 ft/s or 98 m/s².

Although a falling object will continue to accelerate until it is made to stop (when it hits the ground), air resistance will slow down that acceleration. Air resistance is approximately proportional to the square of the velocity, so as the object falls faster, the air resistance increases until it equals the force of gravity. The point at which these forces are equal is called its **terminal velocity**. For instance, a falling baseball reaches 94 miles per hour or 42 meters/second; it would remain at that velocity and no longer accelerate. A penny dropped from a high building will accelerate until it reaches around 230 mph.

The acceleration of the force of gravity on falling bodies is independent of the mass of the falling object. Therefore, a 15-pound weight would fall at the same rate as a 2-pound weight and would hit the ground at the same time if dropped from the same height. In addition, gravity of the mass of an object is also *independent of the velocity* of the object parallel to the ground. For example, two bullets are positioned at the same initial height. One is shot from a gun horizontal to the surface. The other bullet is dropped at the exact same time as the bullet is fired from the gun. Both bullets will hit the ground at the exact same time, if no incidental air resistance or friction exists.

$$F = ma$$

Since *force = mass times acceleration*, the universal gravity equation implies that as objects are attracted and get closer together, the force increases and the acceleration between them also increases.

CLASSICAL MECHANICS

Mechanics is the study of things in motion. **Classical mechanics** involves particles bigger than atoms and slower than light. **Newton's Laws of Motion** form the basis of most of our understanding of things in motion. These three laws are:

#1 Law of Inertia: A particle at rest will stay at rest and a particle in motion will stay in motion until acted upon by an outside force;

#2 Law of Force versus Mass: The rate of change of a particle is directly proportional to its mass and the force that is exerted on it, or

$$F = m \times a \quad \text{or} \quad F = ma$$
(where F is force, m is mass, and a is acceleration);

#3 Law of Action and Reaction: Mutual interactions between bodies produce two forces that are equal in magnitude and opposite in direction; one body exerts a force on the second and the second body exerts an equal force on the first.

The study of classical mechanics is predominately involved with evaluating relationships of motion using equations. When studying relationships in classical mechanics, we usually identify quantities in terms of both their magnitude and direction. These mathematical quantities are called **vectors**. A vector recognizes both the size and direction of the dimension being considered. For example, velocity involves both a speed and a direction of the object. The formula for momentum (p) would then be written as $p = mv$, where p and v are both vector quantities, that is, they represent both a quantity of magnitude (or size) and a direction in which the object is moving. Vectors are identified with either bold letters or an arrow over the letter that indicates direction. It is important to understand how to do vector multiplication in standard physics courses. However, it will not be necessary to practice with vectors in a basic physics course, except to know that this is a standard procedure that may be described in some physics problems you may encounter as you study the material.

The following definitions and equations are those central to the general study of classical mechanics:

— **Work** is the movement of a mass over a distance:

$$\text{Work} = \text{Force} \times \text{Distance}$$
$$W = F \times d \text{ or } W = Fd$$

— **Speed** is the rate of change of an object's distance traveled.

$$s = \frac{d}{t}$$

— **Displacement** measures the change in position of an object, using the starting point and ending point and noting the direction. Consider that displacement may be in some cases the same as distance, but in other cases very different. For instance, if a biker traveled from point A to B to C to D in the following diagram, the distance would equal A + B + C + D. However, the displacement would be 0 since the biker started and ended at the same location (ending with no displacement).

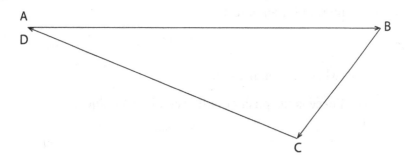

— **Velocity** is the rate of change of displacement; it includes both speed and direction.

$$v = \frac{d}{t}$$

(where v is velocity and has a directional component of plus or minus; d is displacement, t is time)

— **Friction** is the rubbing force that acts against motion between two touching surfaces. Friction between any two objects depends upon the surface attractions or the roughness or smoothness of the two surfaces. The measurement of the amount of friction between any two given objects can be determined experimentally and is designated by the Greek symbol mu, μ, called the coefficient of friction. There are tables that give standard coefficients of friction for some objects that have already been determined experimentally. If μ approaches zero, then there is very little friction. As the value for μ increases, it indicates that there is increasing friction between the objects. For example, tires with worn tread on an icy road would have a low coefficient of friction (μ), and therefore a car with such tires would likely skid. Tires with heavy tread would have a higher coefficient of friction (the addition of sand on the road or chains on the tires would increase this even further) and greatly decrease the likelihood of the car skidding on an icy road.

- **Acceleration** is the rate of change of velocity; it can act in the direction of motion, at an angle, or opposite to the direction of motion.

$$a = \frac{v_2 - v_1}{t_2 - t_1}$$

- **Momentum** is the product of mass and velocity; the quantity of motion for an object.

$$p = mv$$

(where p = momentum)

- **Force** is the push or pull exerted on an object.

$$F = ma$$

$$F = \frac{w}{d}$$

It is also important to note measurements related to mechanics. In science, all measurements are recorded and reported in **metric** units, known as the Système Internationale (International System) or **SI** units:

Mass—measure in kilograms

Length—measured in meters

Time—measured in seconds

Volume—measured in liters

THEORY OF RELATIVITY

Classical mechanics and its resultant laws and equations hold true for systems where we can recognize certain frames of reference. For instance, when a person is traveling on a train going east and walking toward the rear of the train (to the west) while tossing a ball in the air, each of the movements can be described within their own frame of reference whether that be in reference to a point on the ground such as a GPS point (the ball is moving east at x velocity), a spot on the train (the ball is moving west at x velocity), or the person (the ball is moving upward at x velocity).

In 1905, Albert Einstein expanded on the work of Faraday, Lorentz, and Maxwell to propose his **"Special Theory of Relativity."** He stated it in the following two postulates again in 1954:

> 1. ". . . the same laws of electrodynamics and optics will be valid for all frames of reference for which the equations of mechanics hold good. . . . A coordinate system that is moved uniformly and in a straight line relative to an inertial system is likewise an inertial system. By the 'special principle of relativity' is meant the generalization of this definition to include any natural event whatever: thus, every universal law of nature which is valid in relation to a coordinate system C must also be valid as it stands, in relation to a coordinate system C which is in uniform translatory motion relative to C."
>
> 2. "The second principle, on which the special theory of relativity rests, is the 'principle of constant velocity of light in vacuo.' This principle asserts that light in a vacuum always has a definite velocity of propagation (independent of the state of motion of the observer or of the source of the light). The confidence which physicists place in this principle springs from the successes achieved by the electrodynamics of Maxwell and Lorentz." *

In basic terms, Einstein's Special Relativity states that

- the speed of light is a constant,
- the laws of physics are the same in all inertial (non-accelerating) reference frames.

The famous equation that resulted from this theory is

$$E = mc^2$$

where E is energy, m is the mass of an object and c is the speed of light.

What these two postulates logically say is that if you measure the velocity of light c to have a particular value, then no matter which inertial (non-accelerated) reference frame you are moving in, you will always measure it to have the same value; this is an experimental fact. It follows that the velocity of

---

*Albert Einstein (1954). *Ideas and Opinions*. Crown Trade Paperbacks.

light will be the same regardless of the motion of the object against which it is being measured.

Classical mechanics "works" because for objects at speeds not approaching the speed of light and at masses much greater than atomic particles, the equations are appropriate. However, Einstein postulated that the speed of light was the only true constant, and that time and distance would always be made relative in order to maintain the speed of light as a constant. In other words, if a plane is traveling toward a given point at 500 miles per hour and shoots a missile forward at 100 mph, you would presume the missile to be traveling at 600 mph at takeoff. The light that is leaving from the plane's warning beacons is presumably also traveling at 500 miles per hour plus the speed of light when it leaves the plane. However, according to Einstein's special theory of relativity, since the speed of light is always constant, the light is actually traveling at the constant speed for light, not that plus 500 mph.

Einstein proposed several experiments to test his special theory of relativity which required equipment and technology unavailable during his time, but which have now been carried out and verified. The experiments have served to confirm Einstein's proposed theories. The theory of special relativity poses questions as to how distance and time might be distorted by travel at the speed of light or by travel in high gravitational fields, which may mimic the effect of speed of light travel. In addition, the special theory of relativity has also added understanding to the field of *quantum mechanics*—the world of very small objects, namely, subatomic particles. Scientists have long known through experimentation that subatomic particles did not follow the laws of classical mechanics, but rather have their own rules for motion. Quantum mechanics explains these classical discrepancies adequately to allow us to work at the subatomic level.

CHAPTER 9
Energy

CHAPTER 9

ENERGY

ELECTRICITY AND MAGNETISM

Electrical charges consist of **electrons** (with their negative charge) gathered on the surface of an object. When the electrical charges are not moving, it is called *static electricity*. When a positively charged ion attracts the electron, a spark may occur, which is a transfer or discharge of the electrical charge.

Electrical charges may also move through substances that are called **conductors**. A flow of electrons through a conductor is an **electrical current**. Some elements are good conductors (such as metals); others are poor conductors (called **insulators**). Plastic, rubber, glass, and wood are insulators.

An electrical **circuit** is the path that an electric current follows. Every electric circuit has four parts—

1. a *source of charge* (voltage) [a battery, generator, or AC source],
2. a *set of conductors* [wires],
3. a *load* [light, meter, appliance, etc.], and
4. a *switch*.

A **closed circuit** is one that has a continuous path for electron flow (no interruptions). An **open circuit** has no flow of electrons because the pathway is interrupted (by a switch, disconnection, etc.).

Voltage refers to the electromotive force that pushes electrons through the circuit. **Amperage** is the measure of the amount of electron flow or current. **Resistance** is a hindrance to current due to objects that deter the current by their size, shape, or type of conductor.

Series and Parallel Circuits

In a series circuit, there is only one path along which the electrons may flow, moving around the circuit along this single pathway from the positive to the negative pole of the battery (source). The current flows through each component in succession. Linking this group of electric cells in a series means that the voltage of the circuit will be equal to the sum total of the voltages in each cell added together.

In a parallel circuit, the electrical devices are connected to provide two or more paths through which the current may flow. Linking electrical cells in parallel will increase the amperage of the circuit (the current of the circuit will equal the sum of the currents in each cell).

Fig. 9-1 Series and Parallel Circuits.

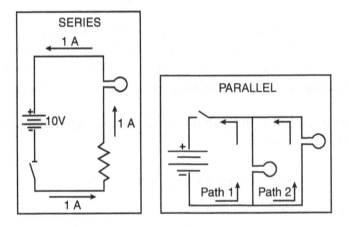

Magnetism

Magnetism is defined as the ability of a substance to produce a magnetic field. The magnetic field acts like point charges (in electricity) producing north (N) and south (S) poles. When magnets are placed in close proximity to each other, the similar poles repel, while opposite poles attract. Magnets can be either permanent or temporary. Permanent magnets are ones that contain natural magnetic ore, such as iron, cobalt or nickel, the three natural elements with magnetic properties. Temporary magnets are ones that can be induced to carry a magnetic field but will not hold the magnetic field permanently. Electromagnets are electrically induced magnets usually created by wrapping coils of wire around an iron core.

WAVES: SOUND AND LIGHT

The study of waves is very important in science. Many things in nature travel in waves, including sound, light, and water. A wave has no mass of its own; it is simply movement within a medium, a disturbance that does not cause the medium itself to move significantly.

There are different types of waves with different types of movement. Light and ocean waves travel in transverse waves. A **transverse wave** causes particles to move up and down while the wave moves forward (perpendicular to the wave motion). Sound and some earthquake waves travel as longitudinal (compression) waves. In a **longitudinal wave** the particles move back and forth but in the same direction as (parallel with) the wave motion.

Fig. 9-2 Longitudinal and Transverse Waves.

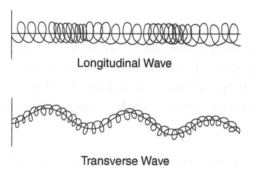

Longitudinal Wave

Transverse Wave

Properties of waves are described with the following terms. The **wavelength** of a wave is defined as the distance from one crest (or top) of a wave to the next crest on the same side. The **frequency** of a wave is the number of wavelengths that pass a point in a second.

Waves also interact with each other in various ways. Wave **interference** can occur between waves. Interference can increase wave amplitude if the crests and troughs of the waves coincide. If the crest and trough of two waves coincide, they can cancel each other or reduce the amplitude. Waves can also be **reflected** when they hit a surface.

The speed of a wave can be described by the equation: $v = f\lambda$, where v = the speed of the wave, f = frequency of the wave, λ = wavelength of the wave.

Fig. 9-3 Properties of Waves.

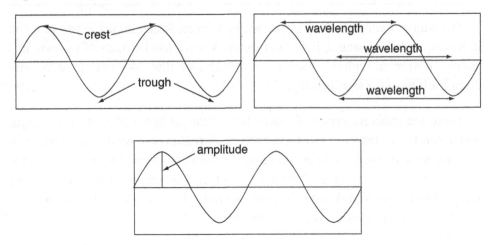

Sound

Sound travels in longitudinal waves. Sound waves are made when an object vibrates and causes air molecules to take on a compression wave motion. This motion is picked up by our eardrums and translated in our brains into sounds. Changes in amplitude account for volume changes, while changes in frequency result in pitch changes.

Anyone who has stood beside a train track and heard the sound of the train whistle or horn has noticed that it sounds the same as the train approaches. As soon as the train passes by, however, the pitch of the horn begins to drop at a regularly reducing pattern as the train moves away from the listener. This phenomenon is known as the "Doppler effect." What is happening here?

As the train approaches, the sound coming to the ear is being pushed by compression waves so that it arrives at the speed of sound, which is the sum of the speed of sound in air plus the speed of the train. Since the sound wave hits the ear drum at the speed of sound or greater, it maintains its regular pitch. However, as the train passes the ear, the speed of the sound wave will begin to decrease proportionately to the ratio of the speed of sound divided by the speed of sound minus the train's speed away from the listener. This results in a lowering of the horn's pitch according to the following equation:

$$f_0 = \left(\frac{v}{v - v_s} \right) f_s$$

where the frequency of the detected sound will be f_o, v is the velocity of sound in air, v_s is the velocity of the train, and f_s is the frequency of the emitted sound wave from the train.

Light

In many situations, light is considered to travel in waves. However, light also has characteristics indicative of a particle, rather than simply a wave.

The Light Spectrum—A spectrum is used to identify the arrangement of the components of a light wave according to wavelength. The visible spectrum of light is the arrangement of the visible wavelengths in a beam of light in the order red orange, yellow, green, blue, violet. Different kinds of light have different spectra. Fluorescent light has sharper bands than incandescent light (light bulb). Sunlight has more blue and violet in its spectrum.

Light travels in straight lines until it hits an object. In order to see an object your eye must intercept a ray of light traveling straight from it. To understand how light affects how things are seen, it is important to understand the ways that light moves by reviewing the following terms:

Diffraction—the bending of a light wave around an obstacle

Reflection—the bouncing of a wave of light off an object

Refraction—the change of direction of a wave as it passes from one medium to another

CHAPTER 10

The Universe

CHAPTER 10

THE UNIVERSE

ASTRONOMY

Galaxies

A **galaxy** is a system that contains stars, star systems (like our solar system), dust, and any other objects within range of the gravitation of its star systems. While all galaxies are massive in scale, the galaxy our solar system resides in, the Milky Way, is thought to be of average size. It is known to contain hundreds of billions of stars including our Sun. Galaxies come in a myriad of shapes and take on various overall colors, but ours is a rotating spiral of milky white. It is theorized that at the center of most galaxies is a black hole. Recent data from space probes and telescopes has given scientists important insights into the nature of black holes. Scientists theorize that these incredibly dense areas of matter arise from remnants of a large star (or the collision of stars) that implodes after a supernova explosion. Once the mass of the star becomes large enough, it collapses under the force of its own gravity. As it collapses, the surface of the star nears an imaginary surface called the "event horizon," as time on the star becomes slower than the time kept by observers far away. When the surface reaches the event horizon, time stands still, and the star can collapse no more. This object contains enough gravity to pull entire star systems into its grasp.

Fig. 10-1 Center of Milky Way Galaxy orbiting a black hole.

Courtesy of NASA

Stars

The core of our Sun, along with most stars, begins with hydrogen. With the core temperature of stars being over 10 million Kelvin, and the tremendous pressures they contain, protons can fuse together to produce helium, gamma ray energy, positrons and neutrinos. These neutrinos are nearly massless and charge-less. They do not interact with other matter very much and flow almost unimpeded throughout the universe. It is speculated that neutrinos produced in the center of the Sun sweep through the average human's body here on Earth at a rate of billions per second.

Stars are huge masses of plasma with colossal amounts of energy and gravity. Many stars are held in by magnetic fields, but particles slip through occasional holes in the fields. The atoms are so hot, the protons, neutrons, and electrons move rapidly and react with each other, thus releasing energy. This energy moves out from stars in electromagnetic waves, producing heat and light. We are too far from most stars to notice their heat, but their light still reaches us.

The stars consist of incredible amounts of matter. Since we know all matter has gravity, stars have very large gravitational forces. Many stars have their own systems, like our solar system with planets, asteroids, and moons orbiting them.

The stars are arrayed in the sky in patterns that ancient people named for various creatures and objects (scales, scorpions, archers, etc.). We call these the zodiac. Travelers have used the zodiac for navigation since biblical times. In addition, Polaris, the North Star, sits directly over our North Pole so it can be used to determine directions on the Earth.

Current scientific theory, including the theory of the **Big Bang**, describes the birth of our universe as a massive explosion occurring approximately fifteen billion years ago. The matter and energy that became the universe was compacted into an infinitely small area. It then began an expansion process at an explosive rate that slowed down over time but that is theorized to still be occurring. The simplest components of matter—protons, neutrons, electrons—came into being and over time condensed into particles of matter. Over hundreds of thousands of years of expansion and cooling, gases condensed and stars were formed, then galaxies, and much later solar systems.

Since the beginning of the universe, stars have been forming and dying. Stars begin as large masses of dust and cosmic gases that collapse together due to gravitational forces. This process of collapse may take millions of years, and during this time the mass is called a nebulae or a proto-star. Over time a process of nuclear fusion of hydrogen (and later helium) atoms begins to power the star at its core, and this will fuel it as a main sequence star for the rest of its life. As the star grows in size, powered by the nuclear fusion reactor at its core, it eventually reaches a red giant phase. Our Sun is a red giant. Eventually, the star will consume its supply of nuclear fuel, and the core will collapse on itself due to its own gravity. At this point it is called a white dwarf. Very large stars may explode into supernovas then coalesce into a dense neutron star or a black hole.

The Sun

Our Sun is a mid-size star that emits heat and light energy. There are stars much larger than our Sun and stars that are smaller. Our Sun is about ten times more massive than the largest planet in the solar system (Jupiter) and 109 times larger than the Earth's diameter.

The Sun releases incredible amounts of energy as protons interact with each other (nuclear fusion). The energy is in the form of heat and light radiation, providing heat and light for life on Earth.

The Solar System

Our solar system consists of all of the common celestial bodies such as planets, moons, asteroids, and various types of space debris that orbit the Sun.

A planet is a celestial body that (a) is in orbit around the Sun, (b) has sufficient mass for its self-gravity to assume a nearly round shape, and (c) has cleared the neighborhood around its orbit.

Fig. 10-2 Approximate relative sizes of planets in relation to the Sun.

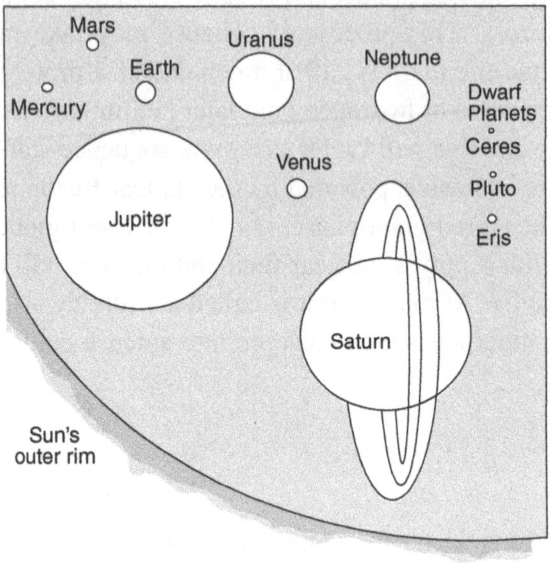

The characteristics of the planets vary greatly because of their size, composition, and distance from the Sun. It is easy to remember the eight planets in order, starting with the closest to the Sun, by learning the following sentence: My Very Educated Mother Just Served Us Noodles. The beginning letter of each word in this sentence is the beginning letter of a planet (Mercury, Venus, Earth, Mars, Jupiter, Saturn, Uranus, and Neptune).

Planet	Type of Planet	Size (to nearest thousand)	Average Distance from Sun (in Miles/Km)	Number of Moons
Mercury	Terrestrial	3,000 mi (5,000 km)	36 million miles (58 million km)	0
Venus	Terrestrial	8,000 mi (12,000 km)	67 million miles (108 million km)	0
Earth	Terrestrial	8,000 mi (13,000 km)	93 million mi. (150 million km)	1
Mars	Terrestrial	4,000 mi (7,000 km)	142 million mi. (228 million km.)	2

Planet	Type of Planet	Size (to nearest thousand)	Average Distance from Sun (in Miles/Km)	Number of Moons
Jupiter	Gas Giant	89,000 mi (143,000 km)	483 million mi. (778 million km.)	67
Saturn	Gas Giant	75,000 mi (120,000 km)	885 million mi. (1,426 million km.)	62
Uranus	Gas Giant	32,000 mi (51,000 km)	1,787 million mi. (2,877 milion km.)	27
Neptune	Gas Giant	31,000 mi (50,000 km)	2,800 million mi. (4.508 million km.)	13

From the table above, you can note several trends about the planets of our solar system. The **terrestrial planets** are so named because they are composed of solid elements. As distance from the Sun increases, so does the size of the planet, and from that information you can infer that the mass of the planet also increases. The planets inside the asteroid belt (which lies outside Mars) are significantly larger and have more mass. The composition of these planets is also different. Being farther from the Sun, they are much colder and are composed of mostly gases with huge atmospheric layers. They are called the **Gas Giants**. Their distance from the Sun means their orbital around the Sun is much longer, making their year much longer. These planets also have much stronger gravitational attraction, since gravity is a function of mass. Therefore, they have captured many more moons into their orbit than the smaller terrestrial planets.

In addition to the planets orbiting our Sun, our solar system also has asteroids and several other minor planets and other space objects. Asteroids are irregular masses of rock and metal that are smaller than planets. There is an asteroid belt between Mars and Jupiter and another beyond the orbit of Neptune.

The outer Solar System is home to comets, masses of frozen gas that travel in orbital patterns. At times, comets travel near the Sun, where the tail of the comet begins to melt and form a stream of gas giving them a long extended look.

Fig. 10-3 Halley's Comet.

Courtesy of NASA

Our calendar year is a measure of the time the Earth takes to orbit the Sun. Each year Earth completes one trip around the Sun in about 365 days. The seasons we experience on Earth are a result of the inclination of the Earth on its axis. That is, if you were to draw a line through the Earth's poles, the line would not be perpendicular with the Earth's orbit around the Sun (see Fig. 10-4).

Fig. 10-4 Tilt of the Earth and seasons.

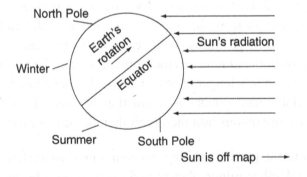

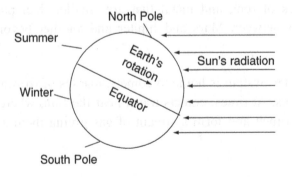

The Earth's axis tilts at an angle of twenty-three and one-half degrees with its orbit. The tilt means that the Northern Hemisphere will be tilted toward the Sun for half the orbit, and the Southern Hemisphere will be tilted for the other half (half-a-year). The hemisphere tilted toward the Sun absorbs more of the solar radiation during that half of the year and experiences summer. The hemisphere tilted away from the Sun experiences winter.

The orbit of the Moon around the Earth occurs approximately every twenty-nine days. This is the origin of our months. A day is the time the Earth takes to rotate one time on its axis. An hour is simply a division of the rotation of the Earth (one day) obtained by dividing the globe into twenty-four time zones.

The Moon

Our view of the Moon is constantly changing as the Moon orbits Earth and as Earth rotates on its axis. Our daily view of the Moon changes as it moves around Earth along its regular orbital path. The changes we see in the shape and location of the Moon are regular in their occurrence because of the regular nature of the rotation and orbit. The Moon orbits around the Earth once approximately every twenty-nine days (one lunar month). The Moon's rotation on its axis is synchronous with its orbit, so we always see the same side of the Moon reflecting the sunlight.

Fig. 10-5 The Phases of the Moon.

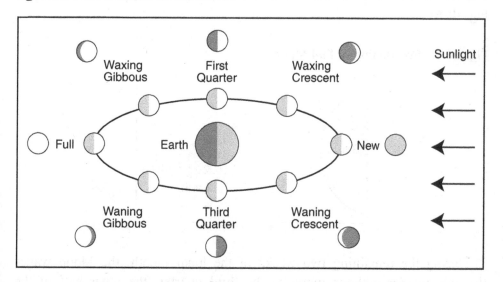

The Moon does not emit any of its own light, but reflects the light from the Sun. Our daily view of the Moon changes during its orbit of the Earth. The phases that we see are a result of the angle between the Earth, Moon, and Sun as viewed by us from Earth.

A *new moon* occurs when the Moon is directly between the Earth and the Sun. At this point, the Moon will rise at about 6:00 a.m. and set at about 6:00 p.m. (on standard, not daylight saving, time). The lighted side of the Moon is toward the Sun, so our view is of the dark side only (we don't see it at all).

Following the new moon, the Moon (continuing its orbit) moves so that we see an increasing portion of the lit side each night. We call this the *waxing crescent moon*.

In about a week, the angle between the Earth, Sun, and Moon is 90°, allowing us to see half its lighted surface, the *first quarter*. This 90° angle means the Moon rises halfway through the day at about noon, and sets at about midnight.

For the following week, we see more of the Moon's surface each night (*waxing gibbous*) until the full moon, which marks the middle of the lunar orbit (and the lunar month).

During the *full moon*, the Earth/Moon/Sun angle is 180°, meaning the Earth is between the Sun and the Moon so we see the entire bright half of its surface. The full moon rises near sundown and sets near sunrise, opposite the Sun (see Fig. 10-6).

Fig. 10-6 Positioning of Full Moon.

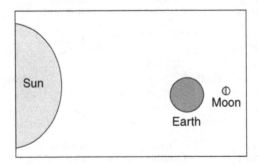

During the remaining two weeks of the lunar month, the Moon wanes through another gibbous moon to the third quarter (the other half of the

Moon's lit surface is visible rising around midnight and setting at noon). It wanes through another crescent moon and on until it returns to the beginning of the orbital—the new moon.

We view the Moon "rising" in the east and "setting" in the west as the Earth rotates on its axis. The Moon's orbit is nearly in the same plane as the orbits of the planets around the Sun, so we view the Moon near that plane in the sky called the *ecliptic*. The tilt of the Earth on its axis means the ecliptic is visible (to those in the northern hemisphere of the Earth) in the southern sky at varying heights through the seasons.

At most times, the angle between the Sun, Moon, and Earth at full and new moon is such that there is no blockage of the view of the Moon or Sun. However, when, during a full moon, the Earth lies directly in the path between the Earth viewer and the Moon, it is called a lunar eclipse, since the Earth blocks the light of the Sun from reaching the Moon for a period of time.

Fig. 10-7 Lunar Eclipse.

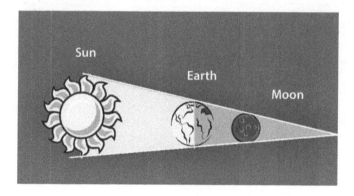

CHAPTER 11
Earth

CHAPTER 11

EARTH

ATMOSPHERE

The atmosphere is comprised of several layers of gases immediately surrounding the Earth. Earth's atmosphere is an essential feature that allows our planet to sustain life. It extends approximately 560 km (350 miles) from the surface of the Earth, though the actual thickness varies from place to place. Our atmosphere makes Earth a habitable planet for people, plants, and animals by absorbing the Sun's energy, recycling and preserving water and chemicals needed for life, and by moderating our weather patterns. The atmosphere is attracted to and maintained by the force of gravity of the Earth.

Without the protection of our atmosphere, Earth would be subjected to the extreme freezing temperatures found in the vacuum of space. The Earth would also be bombarded by dangerous amounts of radiant energy from the Sun. Our atmosphere is specifically formulated to provide us with an environment suited to our needs.

Earth's atmosphere (Fig. 11-1) is made up of about 78% nitrogen, 21% oxygen, slightly less than 1% argon, and the remaining fraction of 1% contains small amounts of other gases (including carbon dioxide, helium, hydrogen, krypton, methane, neon, nitrogen dioxide, nitrous oxide, ozone, sulfur dioxide, water vapor, and xenon).

Fig. 11-1 Atmospheric Composition.

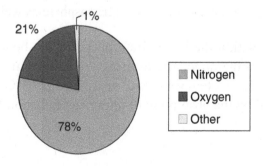

We identify five distinct layers (called strata) of atmosphere: the troposphere, stratosphere, mesosphere, thermosphere, and exosphere. Each layer differs from the others by its temperature, density, and chemical composition. Four transition zones separate the four atmospheric layers: the tropopause, stratopause, mesopause, and thermopause.

The **troposphere** is the atmospheric layer closest to the Earth's surface. It extends to an altitude of approximately 8 to 15 kilometers (5 to 9 miles). The force of gravity is strongest nearest the surface of the Earth, thus the number of gas molecules per area (density) is greatest at the lowest altitudes. In addition, the density of gas molecules decreases as altitude increases. Therefore, the troposphere is the densest atmospheric layer, accounting for most of the mass of the atmosphere.

The troposphere contains 99% of the water vapor found in the atmosphere. Water vapor in the air absorbs solar energy and absorbs heat that radiates back from the Earth's surface. The water vapor concentration within the troposphere is greatest near the equator and lowest at the poles.

Nearly all weather phenomena experienced on Earth are caused by the interactions of gases (including water vapor) within the troposphere. The temperature of the troposphere decreases as altitude increases from about 16°C (60°F) nearest Earth's surface to –65°C (–85°F) at the tropopause. (Temperatures are averages of all the temperatures over the surface of the Earth and across all seasons.) For every kilometer in altitude above the Earth, the temperature drops approximately 6°C within the troposphere.

The troposphere is separated from the next layer (the stratosphere) by the tropopause.

The **stratosphere** is above the troposphere. The stratosphere extends from the tropopause (found approximately 14 km or 9 miles above the Earth's surface) to approximately 48 km (30 miles) above the Earth's surface. The stratosphere contains much less water vapor than the troposphere. The gases of the stratosphere are much less dense than in the troposphere as well.

The temperature within the lower stratosphere (up to about 25 km altitude) is mostly constant. In the upper stratosphere, the temperature rises gradually with increased altitude to a temperature of approximately 3°C. The rising temperatures are caused by the absorption of ultraviolet radiation from the Sun by ozone molecules.

Ozone (O_3) molecules form the ozone layer at the upper ranges of the stratosphere. Ozone molecules absorb solar ultraviolet radiation, which is converted to kinetic energy (heat). This process accounts for the increased temperature levels as altitude increases within the stratosphere. The ozone layer also performs the crucial function of protecting organisms from the harmful effects of too much ultraviolet radiation.

The stratosphere is separated from the next atmospheric layer (the mesosphere) by the stratopause.

The **mesosphere** is the atmospheric layer found at approximately 50–80 km altitude. The mesosphere is characterized by temperatures decreasing with increased altitude from about 3°C at the stratopause to –110°C at 80 km. The mesosphere has a low density of molecules with very little ozone or water vapor. The atmospheric gases of the upper mesosphere separate into layers of gases according to molecular mass. This phenomenon is caused by the weakened effects of gravity on the gas molecules (because of distance from Earth). Lighter (low molecular mass) gases are found at the higher altitudes.

The **mesopause** separates the stratosphere from the next layer, the thermosphere.

The **thermosphere** is found at altitudes of approximately 80 to 480 kilometers. Gas molecules of the thermosphere are widely separated, resulting in very low gas density. The absorption of solar radiation by oxygen molecules in the thermosphere causes the temperature to rise to approximately 1980°C at the upper levels of the thermosphere.

The final layer, the **exosphere** extends from the **thermopause** at approximately 480 km to an altitude of 960 to 1000 kilometers. The exosphere, however, is difficult to define and is more of a transitional area between Earth and space than a distinct layer. The low gravitational forces at this altitude hold only the lightest molecules, mostly hydrogen and helium. Even these are at very low densities.

In addition to the names of the five atmospheric layers, some other terms describe various levels of the atmosphere. The troposphere and tropopause together are sometimes referred to as the "lower atmosphere." The stratosphere and mesosphere are sometimes called the "middle atmosphere," while the thermosphere and exosphere are together known as the "upper atmosphere." Still other scientists call the entire lower area of Earth's atmosphere, embracing the troposphere, mesosphere, and stratosphere together as the "homosphere," and term the thermosphere and exosphere together the "heterosphere."

The **ionosphere** includes portions of the mesosphere and thermosphere. The term *ionosphere* refers to the portion of the atmosphere where ultraviolet radiation causes excitation of atoms resulting in extreme temperatures. Under extreme temperature conditions, electrons are actually separated from the atoms. The highly excited atoms are left with a positive charge. Charged atoms (ions) form layers within the thermosphere. The charged layers are referred to as the ionosphere. The charged particles within the ionosphere deflect some radio signals. While some radio frequencies are not affected by the ionosphere, others are. This reflection of some radio signals causes some radio frequencies (particularly AM transmissions) to be received far from their origination point.

Solar flares create magnetic storms in the thermosphere near Earth's poles. These storms temporarily strip electrons from atoms. When the electrons rejoin the atoms, brilliant light (in green and red) is emitted as they return to their normal state. These lights are called **auroras,** or the Northern and Southern lights.

The total weight of the atmosphere exerts force on the Earth. This force is known as **atmospheric pressure** and can be measured with a **barometer.**

EARTH'S LAYERS

The **geosphere** is the solid or mineral part of the Earth. It consists of layers, from the outer crust down to the inner core separated according to density and temperature. There are two ways to classify the composition of the geosphere:

1. chemically, into crust, mantle, and core, or
2. functionally, into lithosphere and asthenosphere.

Crust, Mantle, and Core

The density of the Earth averages three times that of water. This density varies depending upon the layer of the Earth being considered.

Fig. 11-2 Layers of the Atmosphere. 1: Mt. Everest; 2: Commercial Jet; 3: Fighter Jet; 4: Ozone Layer; 5: Auroras; 6: Space Shuttle.

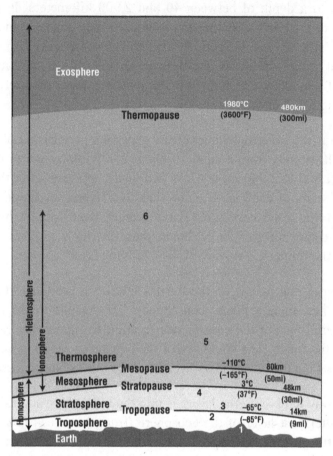

The **crust** is the outermost layer of the geosphere, what we think of as "the Earth." It includes the mountains, valleys, continents, ocean beds, etc. The crust is rich in oxygen, silicon, and aluminum, with lesser amounts of other elements like iron, nickel, etc. It has low density (2.5 to 3.5 gm/cm^3), that allows it to float on the denser mantle. It is brittle and breaks relatively easily. It is made up mostly of sedimentary rocks resting on a base of igneous rocks. Several separate tectonic plates float beneath it on the surface of the mantle.

It is estimated that the entire surface of the Earth can re-form in about 500 million years through erosion and subsequent re-creation through tectonic activity.

The **mantle** is the complex middle layer of the geosphere. It is a broad layer of dense rock and metal oxides that lies between the molten core and the crust and extends to a depth of between 40 and 2,900 kilometers. It accounts for around 82% of the Earth's volume and is thought to be made up mostly of iron, magnesium, silicon, and oxygen. Analysis of seismic waves shows that the material that makes up the mantle behaves as aplastic—a substance with the properties of a solid but which flows under pressure. More precisely, the mantle consists of rigid and plastic zones.

The **core** is the innermost layer of the geosphere, composed of mostly iron and nickel. It extends from a depth of about 2,900 kilometers to nearly 6,400 kilometers (1,800 to 3,900 miles). It is extremely hot, even after thousands (to millions) of years of cooling. The core has two layers, the liquid **outer core,** and the solid iron **inner core** at the Earth's center. Even with the high temperatures at the center (up to 7,500K, hotter than the Sun's surface), this layer is solid due to the immense pressure of the overlying layers.

The material that makes up the Earth's interior is similar to that of liquid steel, but it maintains a fluidic consistency. This results in the migration of Earth's material toward the equator due to angular momentum caused by the spinning of the Earth. The effect is that the equatorial radius is about 21 kilometers larger than the polar radius, which results in its oblate spheroid shape.

Another way to classify the layers is into the lithosphere and asthenosphere. The rigid outermost layer of the geosphere (from the Greek *lithos*—stone) is called the **lithosphere.** The upper layer of the lithosphere is the crust. Beneath the crust is a layer of rigid mantle. The **asthenosphere** is the molten plastic outer mantle of hot silicate rock beneath the lithosphere (from the Greek *asthenos*—devoid of force).

The Discontinuities

The **Mohorovicic** discontinuity or **Moho** is the sharp boundary between the crust and mantle. The **Gutenberg** discontinuity separates the mantle from the core.

The Hydrosphere

The Earth is unique among planets in the solar system because of the large quantities of water that cover its surface. This water is responsible for much of its formation and dynamics. The interconnections between climate and geophysical processes due to hydrology are complex. Two-thirds of the planet's surface is covered by water at an average ocean depth of 3.9 kilometers (12,795 feet). The water area is 3.6×10^8 square kilometers (nearly 140 million square miles).

The Magnetosphere

The magnetic field that surrounds the Earth extends thousands of miles into outer space. It mimics a bar magnet with the poles being approximately located at the north and south poles and a neutral area at the equator. The poles shift slightly over time. The power of the Earth's magnet is not very strong, being approximately .5 Gauss at the surface of the Earth.

The location of the North Magnetic Pole was determined in 1996 by a Canadian expedition and certified by magnetometer and theodolite at 78°35.7′N 104°11.9′W. In 2005, the location was at 82.7°N 114.4°W, slightly west of Ellesmere Island in Northern Canada. It is clear that the magnetic field moves over time.

The Geologic Column

The estimated age of Earth is determined by radiometric dating and geological estimates to be about 4.6 billion years. The geologic column represents the layers of the Earth near the surface that contain fossils and various types of sediment. The geologic column is currently used as a tool of evolutionary uniformitarian scientists to display a "progression" of life from simple species in the deepest (oldest) layers to more complex species in the layers closer to the surface (younger). The layers are categorized into Eras, Periods, and Epochs. Fig. 11-3 represents the general arrangement of the geologic column.

Fig. 11-3 Geologic Column.

Geologic Time Scale

Eon	Era	Period	Epoch	Age (my)
Phanerozoic (Visible Life)	Cenozoic (Recent Life) (Age of Mammals)	Quaternary	Holocene	0.01
			Pleistocene	1.6
		Tertiary	Pliocene	5.3
			Miocene	23.7
			Oligocene	36.6
			Eocene	57.8
			Paleocene	66.4
	Mesozoic (Middle Life) (Age of Reptiles)	Cretaceous		144
		Jurassic		208
		Triassic		245
	Paleozoic (Ancient Life)	Permian		286
		Pennsylvanian		320
		Mississippian		360
		Devonian		408
		Silurian		438
		Ordovician		505
		Cambrian		570
Proterozoic (Early Life)				
Archean	——— Oldest Known Life ———			2500
	——— Oldest Known Rocks ———			3900
Hadean	——— Age of Earth ———			4600

PRACTICE TEST 1
CLEP Natural Sciences

Also available at the REA Study Center (*www.rea.com/studycenter*)

This practice test is also available online at the REA Study Center. The CLEP Natural Sciences test is only offered as a computer-based exam; therefore, we recommend that you take the online version of the practice test to receive these added benefits:

- **Timed testing conditions** – Gauge how much time you can spend on each question.
- **Automatic scoring** – Find out how you did on the test, instantly.
- **On-screen detailed explanations of answers** – Learn not just the correct answer, but also why the other answer choices are incorrect.
- **Diagnostic score reports** – Pinpoint where you're strongest and where you need to focus your study.

PRACTICE TEST 1

CLEP Natural Sciences

(Answer sheets appear in the back of the book.)

TIME: 90 Minutes
        120 Questions

DIRECTIONS: Each of the following groups of questions consists of five lettered terms followed by a list of numbered phrases or sentences. For each numbered phrase or sentence, select the one choice that is most clearly related to it. Each choice may be used once, more than once, or not at all. Fill in the corresponding oval on the answer sheet.

Questions 1–5

  (A) Secretory vesicle
  (B) Smooth endoplasmic reticulum
  (C) Microvilli
  (D) Nucleolus
  (E) Nuclear pore

1. ____ Communication channel between the cytoplasm and nucleoplasm

2. ____ Extensions that provide extra surface area for absorption

3. ____ Contain digestive enzymes

4. ____ Packets that carry substances (hormones, fats, etc.) synthesized within the cell

5. ____ Network of membranes that deliver lipids and proteins throughout the cytoplasm

Questions 6–8

(A) KCl
(B) NaF
(C) Cu
(D) SiC
(E) $HC_2H_3O_2$

6. _____ Atoms are held together with network covalent attraction.

7. _____ Atoms are held together with covalent attraction.

8. _____ Atoms are held together with metallic attraction.

Questions 9–13

(A) Nodes
(B) Nonvascular plants
(C) Angiosperms
(D) Gymnosperms
(E) Lateral buds

9. _____ Bryophytes

10. _____ Flowering plants

11. _____ Shoots and roots

12. _____ Produce seeds without flowers

13. _____ Conifers and cycads

Questions 14–16

(A) Radiation
(B) Convection
(C) Irradiation
(D) Conduction
(E) Diffusion

14. _____ An athlete with a sore shoulder places a warm compress on it to transfer energy to soothe the muscle.

15. _____ On a cold February morning, a blower system in a car warms up after several minutes and blows air through vents in the floor, dashboard, and windshield. Eventually, the driver is able to unbutton his coat and stay warm when the outside temperature is still 23°F.

16. _____ Getting ready for a fall cruise inspires a young lady to spend a couple of weeks going to a local spa and reclining under a tanning lamp. However, such practices might result in dangerous overexposure to ultraviolet rays that can lead to cancer or premature aging of the skin.

Questions 17–19

(A) H_2
(B) $KMnO_4$
(C) MgO
(D) KCl
(E) Fe_2O_3

17. _____ This substance is a very strong oxidizing agent.

18. _____ The metal in this substance has an oxidation number of +2.

19. _____ The oxidation potential of this substance is zero.

Questions 20–24

(A) Natality
(B) J-curve
(C) Population rate of growth
(D) Mortality
(E) S-curve

20. _____ Exponential population growth curve; population growth accelerates

21. _____ Birth rate minus death rate

22. _____ Death rate

23. _____ Birth rate

24. _____ Logistic population growth curve; population growth accelerates and then slows down because of limits

DIRECTIONS: Each of the questions or incomplete statements below is followed by five possible answers or completions. Select the best choice in each case and fill in the corresponding oval on the answer sheet.

25. Starting from rest, the distance a freely falling object will fall in 0.5 seconds is about

    (A) 0.5 m.
    (B) 1.0 m.
    (C) 1.25 m.
    (D) 5.0 m.
    (E) 5.25 m.

26. Which of the following mechanisms of evolution can produce genetic changes in a community that did not originally exist in the gene pool?

    (A) genetic drift
    (B) allopatric speciation
    (C) sympatric speciation
    (D) mutation
    (E) natural selection

27. Which of the following statements is NOT correct when cells undergo meiosis?

    (A) Meiosis ensures that the chromosome number remains constant generation after generation.
    (B) Meiosis ensures that each generation has a different genetic makeup than the previous one.
    (C) Meiosis ensures that each newly-formed daughter cell receives the same number and kinds of chromosomes.
    (D) Meiosis results in four daughter cells.
    (E) Meiosis occurs in the production of egg and sperm cells in animals.

28. Which of the following classification groups is known for radial symmetry in most of its species?

    (A) felidae
    (B) echinodermata
    (C) platyhelminthes
    (D) chondrichthyes
    (E) Astralopithicus

29. Which of the following processes results in genetic variation?

    (A) interphase
    (B) meiosis
    (C) metaphase
    (D) prophase
    (E) mitosis

30. The evolution of plant species is considered to have begun with

    (A) autotrophic prokaryotic cells
    (B) heterotrophic eukaryotic cells
    (C) aerobic eukaryotic cells
    (D) pre-nucleic eukaryotic cells
    (E) aerobic prokaryotic cells

31. What type of a wave is a sound wave?

   (A) Compression
   (B) Transverses
   (C) Inverse
   (D) Converse
   (E) Convex

32. Salinity is a measure of dissolved solids in water. On average, 1,000 g of typical sea water contains 35 g of salt. The level of salinity is below average in areas where large amounts of freshwater enter the ocean and above average in hot, arid climates. The Mediterranean and the Red seas are adjacent to deserts. The water in these two seas would be expected to have

   (A) above-average salinity
   (B) below-average salinity
   (C) average salinity
   (D) no salt content
   (E) cold temperatures

33. The discoveries of Galileo Galilei (1564–1642) and Isaac Newton (1642–1727) precipitated the scientific revolution of the seventeenth century. Stressing the use of detailed measurements during experimentation enabled them to frame several universal laws of nature and to overthrow many of Aristotle's (384–322 B.C.E.) erroneous ideas about motion, which were based on sheer reasoning alone. One of these universal laws is now known as the law of inertia or Newton's first law of motion. According to this law, objects in motion tend to stay in motion and objects at rest tend to stay at rest unless acted upon by an external force. The more mass (inertia) an object has, the more resistance it offers to changes in its state of motion. According to the law of inertia, which of the following would offer the greatest resistance to a change in its motion?

   (A) A pellet of lead shot
   (B) A golf ball
   (C) A large watermelon
   (D) A feather
   (E) A sheet of notebook paper

34. The site of transfer for nutrients, water, and waste between a mammalian mother and embryo is the

    (A) yolk sac membrane
    (B) uterus
    (C) placenta
    (D) umbilical membrane
    (E) ovary

35. The process of genetic inheritance was first investigated and explained by

    (A) Robert Hooke
    (B) Dmitri Mendeleev
    (C) Theodor Schwann
    (D) Matthias Schleiden
    (E) Gregor Mendel

36. The first cells to evolve on Earth were most likely NOT

    (A) anaerobic
    (B) specialized
    (C) prokaryotic
    (D) aquatic
    (E) small

37. Which of the following best represents the sequence of human evolution?

    (A) *Australopithecus afarensis*, Cro-Magnon, *Homo erectus*, *Homo sapiens*, Modern man
    (B) Cro-Magnon, *Australopithecus afarensis*, *Homo erectus*, *Homo sapiens*
    (C) *Homo sapiens*, *Australopithecus afarensis*, *Homo erectus*, Cro-Magnon
    (D) *Australopithecus afarensis*, *Homo erectus*, Cro-Magnon, Modern man
    (E) Cro-Magnon, *Homo erectus*, *Australopithecus afarensis*, Modern man

38. Which of the following demonstrates pi bonding?

    (A) OH^2
    (B) H^1
    (C) C_2H_2
    (D) H_2S
    (E) KCl

39. The Sun crosses the celestial equator going north on March 21. This is known as the

    (A) solstice
    (B) lunar eclipse
    (C) solar eclipse
    (D) spring equinox
    (E) autumnal equinox

40. Cells of eukaryotes have all of the following EXCEPT

    (A) membraned organelles
    (B) DNA organized into chromosomes
    (C) a nucleus
    (D) DNA floating free in the cytoplasm
    (E) ribosomes

41. A small non-protein molecule such as iron that works with enzymes to promote catalysis is known as

    (A) a protein
    (B) an inorganic cofactor
    (C) a coenzyme
    (D) a hormone
    (E) a carbohydrate

42. Which of the following will NOT inhibit enzymatic reactions?

    (A) Temperature
    (B) pH level
    (C) Particular chemical agents
    (D) Lack of substrate
    (E) Large amount of enzyme

For questions 43–46, identify the answer that corresponds with the correct functional group.

(A) R—C(H)(H)—OH

(B) R—C(=O)—H

(C) R—C(=O)—O—C(H)(H)—R

(D) R—C(=O)—R

(E) R—C(=O)—OH

43. _____ Ketone

44. _____ Alcohol

45. _____ Ester

46. _____ Aldehyde

47. The conversion of light energy into chemical energy is accomplished by

(A) catabolism
(B) oxidative phosphorylation
(C) metabolism
(D) protein synthesis
(E) photosynthesis

48. Which of the following terms is defined as the increase in the statistical disorder of a physical system?

   (A) Enthalpy
   (B) Entropy
   (C) Epathy
   (D) Euprophy
   (E) Empathy

49. Which of the following evolutionary developments is out of sequence?

   (A) Formation of coacervates occurred in primordial seas.
   (B) Development of eukaryotic cells took place.
   (C) The process of photosynthesis was developed within single cells.
   (D) Amphibians diversified into birds, then mammals.
   (E) A branch of mammals developed into primates, the direct ancestors of man.

50. Energy flows through the food chain from

   (A) producers to consumers to decomposers
   (B) producers to secondary consumers to primary consumers
   (C) decomposers to consumers to producers
   (D) secondary consumers to producers
   (E) consumers to producers

51. If the Moon completely covers the sun as seen by an earthbound observer, there is a

   (A) total lunar eclipse
   (B) total solar eclipse
   (C) partial lunar eclipse
   (D) partial solar eclipse
   (E) parallax

52. The projection of the Earth's axis on the sky is the

   (A) celestial equator
   (B) zenith
   (C) celestial poles
   (D) ecliptic
   (E) zodiac

53. The process that releases energy for use by the cell is known as

    (A) photosynthesis
    (B) aerobic metabolism
    (C) anaerobic metabolism
    (D) cellular respiration
    (E) dark reaction

54. Which of the following is NOT a step in the translation portion of protein synthesis?

    (A) tRNA anticodons line up with corresponding mRNA codons
    (B) a ribosome attaches to the start codon on mRNA; ribosome adds tRNA whose anticodons complement the next codon on the mRNA string; and the process repeats to locate sequential amino acids
    (C) Ribosomal enzymes also link the sequential amino acids into a protein chain
    (D) As the protein chain is forming, rRNA moves along the sequence adding amino acids designated by codons on the mRNA
    (E) Terminating codon stops the synthesis process and releases the newly formed protein

55. Which of the following most closely represents the sequence leading to human evolution?

    (A) Coacervates, eukaryotes, plants, fish, amphibians, mammals, primates, man
    (B) Eukaryotes, plants, amphibians, fish, mammals, primates, man
    (C) Eukaryotes, coacervates, plants, fish, amphibians, mammals, primates, man
    (D) Coacervates, eukaryotes, plants, fish, amphibians, primates, mammals, man
    (E) Plants, eukaryotes, coacervates, fish, amphibians, mammals, primates, man

56. Water molecules are attracted to each other due to which of the following?

    (A) Polarity; partial positive charge near hydrogen atoms; partial negative charge near oxygen atoms
    (B) Inert properties of hydrogen and oxygen
    (C) Ionic bonds between hydrogen and oxygen
    (D) The crystal structure of ice
    (E) Brownian motion of hydrogen and oxygen atoms

57. The voltage across the terminal of a circuit containing a resistor is 12 V, and it contains a resistor connected to the circuit which reads 8 ohms. What is the amperage of the service?

   (A) 1.5 A
   (B) 48 A
   (C) 0.67 A
   (D) 15 A
   (E) 96 A

58. The percentage of Earth's surface that is water is

   (A) 10 percent
   (B) 30 percent
   (C) 50 percent
   (D) 70 percent
   (E) 90 percent

59. Traits that are produced by the expression of more than one set of genes are known as

   (A) polygenic traits
   (B) autosomal traits
   (C) sex-limited traits
   (D) monohybrid traits
   (E) dihybrid crosses

60. The major driving force of the evolution of species is known as

   (A) the Oparin Theory
   (B) natural selection
   (C) environmental determinism
   (D) Hardy-Weinberg Equilibrium
   (E) genetic drift

61. Which of the following items found on a bicycle is NOT a simple machine?

   (A) tire
   (B) pedal mechanism
   (C) rear-wheel gear mechanism
   (D) horn
   (E) kickstand

62. Which of the following sets of quantum numbers (listed in order of n, l, ml, ms) describe the highest energy valence electron of nitrogen in its ground state?

    (A) 2, 0, 0, 1½
    (B) 2, 1, 1, 2½
    (C) 2, 1, 1, 1½
    (D) 2, 1, 2l, 2½
    (E) 2, 1, 2l, 1½

63. The change of state from liquid to solid or solid to liquid involves a phase where the temperature remains constant. This phase is known as the

    (A) transition phase
    (B) heat of fusion
    (C) heat of fission
    (D) specific heat
    (E) equilibrium

64. Which of the following solid crystals has, on average, one atom per cubic unit cell?

    (A) Face-centered
    (B) Body-centered
    (C) Rhombic
    (D) All cubic crystals
    (E) Simple cubic crystals only

65. Which of the following is part of the alimentary canal?

    (A) Artery
    (B) Sinus
    (C) Vagus nerve
    (D) Bronchus
    (E) Esophagus

66. Which of the following experimental evidence was NOT considered supportive of the Oparin Hypothesis?

(A) Amino acids can be produced in the laboratory by exposing simple inorganic molecules to electrical charge.
(B) Guanine can be formed in the laboratory by thermal polymerization of amino acids.
(C) Ultraviolet light induces the formation of dipeptides from amino acids in laboratory experiments.
(D) In the laboratory, proteins are not useful as catalysts, indicating that early proteins were stable.
(E) Phosphoric acid increases the yield of polymers in the laboratory, simulating the role of ATP in protein synthesis.

Questions 67–71

In snapdragons, a red flower crossed with a white flower produces a pink flower. In this illustration, R stands for red and W represents white. The Punnett square for a cross between a white snapdragon and a red snapdragon is shown here:

	R	R
W	RW	RW
W	RW	RW

67. The cross illustrated in this Punnett square is an example of

(A) a sex-linked trait
(B) multiple alleles
(C) incomplete dominance
(D) a dihybrid cross
(E) complete dominance

68. The symbol RW represents which of the following?

(A) The allele for red
(B) The genotype for pink
(C) The phenotype for pink
(D) The allele for white
(E) The genotype for white

69. In this cross, both parents have genotypes that are

    (A) heterozygous for color
    (B) homozygous for color
    (C) recessive for pink
    (D) dominant for pink
    (E) dominant for white

70. Which of the following statements about this cross must be true?

    (A) Both parents of the red snapdragon must have had the genotype RR.
    (B) One of the parents of the red snapdragon must have had the genotype RR.
    (C) Both parents of the red snapdragon must have been pink.
    (D) Neither parent of the red snapdragon could be white.
    (E) One parent of the red snapdragon could have been white.

71. If two of the heterozygous offspring (RW) of this cross are bred, what will be the ratio of phenotypes of the offspring?

    (A) 0 red: 4 pink: 0 white
    (B) 2 red: 2 pink: 0 white
    (C) 1 red: 1 pink: 1 white
    (D) 2 red: 0 pink: 2 white
    (E) 1 red: 2 pink: 1 white

72. The physical place where a particular organism lives is called a

    (A) niche
    (B) biosphere
    (C) lithosphere
    (D) habitat
    (E) ecosystem

73. What is the most likely explanation for the fact that a sample of solid nickel is attracted into a magnetic field, but a sample of solid zinc chloride is not?

    (A) There are unpaired outer electrons in nickel.
    (B) There is some iron mixed in with the nickel.
    (C) There are unpaired outer electrons in zinc.
    (D) The presence of chlorine keeps zinc from being attracted to the magnet.
    (E) Nickel does not produce a magnetic field to oppose the one that is attracting it.

74. The specific heat of water is 4.2 J/g °C. What mass of water will be heated by 10°C by 840J?

   (A) 0.5 g
   (B) 10 g
   (C) 20 g
   (D) 4.2 g
   (E) 840 g

75. A light-year represents the

   (A) total amount of light energy that travels past a point on Earth in one year
   (B) total distance that an object travels in space at the speed of light in one year
   (C) total amount of time that a photon of light travels in order to reach a star, planet, or other extraterrestrial object
   (D) speed at which light travels in a year
   (E) speed of a photon in a vacuum when it collides with an x-ray

76. The half-life of C 14 is 5,730 years. A piece of cypress is measured to have ¾ of the carbon to be C 12 and ¼ C 14. How many years old is this piece of wood likely to be?

   (A) 5,730
   (B) 2,865
   (C) 4,297
   (D) 11,460
   (E) 1,435

77. Igneous rock forms are commonly called

   (A) marble
   (B) limestone
   (C) cement
   (D) granite
   (E) sandstone

78. Which of the following statements is true?

    (A) Despite the contraction of space, galaxies appear to be static relative to each other when observed.
    (B) The expansion of space makes galaxies appear to be moving apart, causing the color of their spectral lines to shift when observed.
    (C) Galaxies themselves are moving apart from each other.
    (D) Hubble's law states that the redshift in light coming from a distant galaxy is inversely proportional to its distance from Earth.
    (E) The actual motion of galaxies is questionable due to variations in Hubble's constant.

79. All of the following are major structural regions of roots EXCEPT the

    (A) meristematic region
    (B) elongation region
    (C) root cap
    (D) epistematic region
    (E) maturation region

80. When a yellow pea plant is crossed with a green pea plant, all the offspring are yellow. The law that best explains this is the law of

    (A) intolerance
    (B) dominance
    (C) interference
    (D) relative genes
    (E) codominance

81. Which of the following statements is true about electrons?

    (A) Electrons have a positive charge.
    (B) Electrons have less mass than protons and neutrons.
    (C) Electrons are found within the nucleus of atoms.
    (D) The number of electrons is equal to the number of protons in an ion.
    (E) An atom's valence number is the number of electrons in its lowest energy level.

82. What is the energy-generating mechanism of the stars, including the Sun?

    (A) Fission
    (B) Fusion
    (C) Spontaneous generation
    (D) Combustion
    (E) Stellar explosion

83. When measuring the flow of heat in a system, a natural process that starts at one equilibrium state and flows to another will go in what direction for an irreversible process that is impacted by the entropy of the system plus the environment?

    (A) Increase
    (B) Decrease
    (C) Remain static
    (D) Fluctuate between increases and decreases in a constant pattern
    (E) Increase steadily and then drop off, similar to the conservation of momentum in a closed system

84. What is the resistance of an electric can opener if it takes a current of 10 A when plugged into a 120 V service?

    (A) 1,200 ohms
    (B) 0.083 ohms
    (C) 130 ohms
    (D) 12 ohms
    (E) 1.2 ohms

85. Photosynthesis would NOT proceed without which of the following that allow moisture and gases to pass in and out of the leaf?

    (A) Surface hairs
    (B) Stomata
    (C) Cuticles
    (D) Epidermal cells
    (E) Cilia

86. In ferns, the individual we generally recognize as an adult fern is really which structure?

    (A) A mature gametophyte
    (B) A prothallus
    (C) A mature sporophyte
    (D) A young sporophyte
    (E) A young gametophyte

87. Which of the following is the least polar molecule?

    (A) H_2
    (B) H_2O
    (C) H_2S
    (D) C_2H_2
    (E) NaH

88. Which state of matter does NOT have a definite shape or a definite volume?

    (A) Solid
    (B) Liquid
    (C) Gas
    (D) Plasma
    (E) Both (C) and (D)

89. A person who has been exercising vigorously begins to sweat and breathe rapidly. These reactions are involuntary responses known as

    (A) fight or flight responses
    (B) homeostatic mechanisms
    (C) equilibrium responses
    (D) stimulus receptors
    (E) hormone reactions

90. The nervous system is an integrated circuit with many functions. Which of the following parts of the nervous system are matched with the wrong function?

    (A) Forebrain — controls olfactory lobes (smell)
    (B) Cerebrum — controls the function of involuntary muscles
    (C) Hypothalamus — controls hunger and thirst
    (D) Cerebellum — controls balance and muscle coordination
    (E) Midbrain — contains optic lobes and controls sight

91. Fog is

    (A) the same as smog
    (B) caused when cold air moves over warm air
    (C) a collection of minute water droplets
    (D) associated with a tornado
    (E) Both (C) and (D)

92. Land and sea (or lake) breezes form because of

    (A) uneven heating of coastal environments
    (B) the pressure gradient force
    (C) the difference in temperature between land and water surfaces
    (D) solar radiation
    (E) a variety of factors involving temperature, pressure, and geographical components

93. An electric shock can restart a heart that has stopped beating. Which of the following statements is a valid reason for this fact?

    (A) Electric shock stimulates the nervous system.
    (B) Smooth muscle is affected by electric shock.
    (C) The electric shock pushes blood through the stopped heart.
    (D) Cardiac muscle of the heart responds to the electric shock.
    (E) Electric shock causes air to enter the lungs.

94. When a hamburger is consumed by an individual, it passes through all of the following organs EXCEPT the

    (A) mouth
    (B) esophagus
    (C) salivary glands
    (D) stomach
    (E) small intestine

95. Each ecosystem can support a certain number of organisms—a number usually designated by the letter K and known as

    (A) ecodensity
    (B) population
    (C) carrying capacity
    (D) a community
    (E) the biosphere

96. Which of the following parts of an atom is NOT a subatomic particle?

    (A) A quark
    (B) A boson
    (C) A neutrino
    (D) An electron
    (E) A positron

97. The function of the gallbladder and pancreas is to aid digestion by producing digestive enzymes and secreting them into the

    (A) small intestine
    (B) large intestine
    (C) stomach
    (D) esophagus
    (E) mouth

98. Many insects have special respiratory organs known as

    (A) spiracles
    (B) alveoli
    (C) cephalothorax
    (D) lungs
    (E) gills

99. Most of Earth's photosynthesis takes place in which one of the following biomes?

   (A) Oceans
   (B) Tundra
   (C) Deciduous forests
   (D) Deserts
   (E) Tropical rainforests

100. Which of the following chemical equations represents a replacement reaction?

   (A) $A + C \rightarrow AC + B$
   (B) $A + B \rightarrow AB$
   (C) $AB + C \rightarrow AC + B$
   (D) $AB \rightarrow A + B$
   (E) $AB + CD \rightarrow A + B + C + D$

101. The study of the interaction of organisms with their living space is known as

   (A) environmentalism
   (B) habitology
   (C) zoology
   (D) ecology
   (E) paleontology

102.. What is the name of a distinct group of individuals that are able to mate and produce viable offspring?

   (A) A class
   (B) A community
   (C) A phylum
   (D) A family
   (E) A species

103. Which of the following is true concerning Einstein's theory of relativity?

    (A) As energy increases, the speed of light increases and mass is constant.
    (B) As energy increases, mass increases and the speed of light is constant.
    (C) As energy increases, the speed of light decreases and mass is constant.
    (D) As energy increases, mass decreases and the speed of light is constant.
    (E) As energy increases, the speed of light and mass will increase.

104. Which of the following is an autotroph?

    (A) *E. coli* bacteria
    (B) A Portuguese man-of-war
    (C) A portobello mushroom
    (D) An asparagus fern
    (E) A perch

105. The Michelson-Morley experiment in 1887 proved that the speed of light

    (A) is the same in all directions
    (B) can slow down in a vacuum
    (C) is slower in air than in liquid
    (D) is different on the Moon
    (E) continuously speeds up in a vacuum

106. Which of the following is NOT a fundamental geological principle?

    (A) The principle of original horizontality
    (B) The principle of supererogation
    (C) The principle of lateral continuity
    (D) The principle of fossil succession
    (E) The principle of uniformitarianism

107. Astronomers observing the redshift of light from a faraway star means that the star is

    (A) moving away from us
    (B) moving toward us
    (C) ready to explode
    (D) a dwarf star
    (E) a giant nova

108. The three most recent geological eras are

   (A) Paleozoic, Proterozoic, Archean
   (B) Mesozoic, Paleozoic, Cenozoic
   (C) Cenozoic, Mesozoic, Paleozoic
   (D) Cenozoic, Archean, Paleozoic
   (E) Recent, Tertiary, Quaternary

109. A form of symbiosis in which one species is benefited while the other is harmed is called

   (A) parasitism
   (B) mutualism
   (C) amensalism
   (D) altruism
   (E) commensalism

110. An axis is

   (A) a connecting line at the equator
   (B) a connecting line between the poles
   (C) a connecting line between orbits
   (D) a pole
   (E) a connecting line at the tropical zones

111. In order of magnitude, how many galaxies are there in the universe?

   (A) Hundreds
   (B) Thousands
   (C) Millions
   (D) Billions
   (E) Trillions

112. An astronaut is traveling in a spacecraft that is slowing down. To the astronaut inside the spacecraft, the apparent force inside the craft is directed

   (A) backward
   (B) forward
   (C) sideways
   (D) vertically only
   (E) nowhere because there is no force

113. The cells of which of the following organisms are most likely to be prokaryotic?

    (A) Mold
    (B) Seaweed
    (C) Blue-green algae
    (D) Fern
    (E) Hydra

114. The name for the imaginary object that centers on and surrounds Earth by which background stars are projected is called a(n)

    (A) geodesic dome
    (B) celestial sphere
    (C) stellar ball
    (D) Cartesian coordinate system
    (E) equatorial coordinate system

115. Why is stratospheric ozone depletion (destruction of the ozone layer) a serious concern?

    (A) It is a major cause of the "greenhouse effect."
    (B) It will increase the amount of ultraviolet radiation reaching the ground.
    (C) It causes acid rain.
    (D) It is really nothing to worry about.
    (E) It leads to global warming as more radiation enters through the hole in the atmosphere.

116. Mountain and valley breezes form because of

    (A) gravity and the pressure gradient force
    (B) the pressure gradient force and heating
    (C) gravity and heating
    (D) the pressure gradient force and the Coriolis force
    (E) high pressure forming near the top of the mountain due to warm air expansion forcing air up the mountain from the valley below

117. As you go down the periodic table and to the left, which of the following traits increases?

    (A) Atomic radius
    (B) Electronegativity
    (C) Electron affinity
    (D) Ionization energy
    (E) Acidity of the oxides

118. According to Newton's laws of motion, the greater the mass of an object, the greater the force necessary to change its

    (A) position
    (B) force
    (C) state of motion
    (D) shape
    (E) density

119. The top or peak of a sine wave is called the

    (A) crest
    (B) trough
    (C) amplitude
    (D) period
    (E) frequency

120. A wave does not carry along the medium through which it travels. Thus, it follows that

    (A) molecules of water in the ocean are pushed to shore by waves
    (B) the ocean's water molecules are thoroughly mixed each day by waves
    (C) debris in the ocean is washed ashore by waves
    (D) individual water molecules do not travel toward shore, but wave peaks do
    (E) waves move water, not swimmers

PRACTICE TEST 1

Answer Key

1. (E)	25. (D)	49. (C)	73. (A)	97. (A)
2. (C)	26. (D)	50. (A)	74. (C)	98. (A)
3. (C)	27. (C)	51. (B)	75. (B)	99. (A)
4. (A)	28. (B)	52. (C)	76. (D)	100. (C)
5. (B)	29. (B)	53. (D)	77. (D)	101. (D)
6. (D)	30. (E)	54. (A)	78. (B)	102. (E)
7. (E)	31. (A)	55. (A)	79. (D)	103. (B)
8. (C)	32. (A)	56. (A)	80. (B)	104. (D)
9. (B)	33. (C)	57. (A)	81. (B)	105. (A)
10. (C)	34. (C)	58. (D)	82. (B)	106. (B)
11. (C)	35. (E)	59. (A)	83. (A)	107. (A)
12. (D)	36. (B)	60. (B)	84. (D)	108. (C)
13. (D)	37. (D)	61. (D)	85. (B)	109. (A)
14. (D)	38. (C)	62. (C)	86. (C)	110. (B)
15. (B)	39. (D)	63. (B)	87. (A)	111. (D)
16. (A)	40. (D)	64. (E)	88. (E)	112. (B)
17. (B)	41. (B)	65. (E)	89. (B)	113. (C)
18. (C)	42. (E)	66. (D)	90. (B)	114. (B)
19. (A)	43. (D)	67. (C)	91. (C)	115. (B)
20. (B)	44. (A)	68. (B)	92. (E)	116. (C)
21. (C)	45. (C)	69. (B)	93. (D)	117. (A)
22. (D)	46. (B)	70. (D)	94. (C)	118. (C)
23. (A)	47. (E)	71. (E)	95. (C)	119. (A)
24. (E)	48. (B)	72. (D)	96. (D)	120. (D)

PRACTICE TEST 1

Detailed Explanations of Answers

1. **(E)** Nuclear pores are holes in the nuclear membrane where the double nuclear membrane fuses together, forming a break or hole, allowing the selective intake and excretion of molecules to or from the nucleus. Thus, nuclear pores are the channel of communication between the cytoplasm and the nucleoplasm.

2. **(C)** Microvilli are filaments that extend from the cell membrane, particularly in cells that are involved in absorption (such as in the intestine). These filaments increase the surface area of the cell membrane, thus increasing the area available to absorb nutrients.

3. **(C)** Microvilli also contain enzymes that are involved in digesting certain types of nutrients.

4. **(A)** Secretory vesicles are packets of material packaged by either the Golgi apparatus or the endoplasmic reticulum. The secretory vesicle carries the substance produced within the cell to the cell membrane. The vesicle membrane fuses with the cell membrane, allowing the substance to escape the cell.

5. **(B)** The smooth endoplasmic reticulum is a network of continuous membranous channels that connect the cell membrane with the nuclear membrane and is responsible for the delivery of lipids and proteins to certain areas within the cytoplasm. The smooth endoplasmic reticulum lacks attached ribosomes.

6. **(D)** Silicon carbide (SiC) is held together with network covalent bonds, which creates a crystal entirely from covalent bonds and confers unusual strength (and a high melting point) upon the crystal.

7. **(E)** Acetic acid is a molecular compound that is held together with covalent bonds, in which electrons are shared between atoms, so that both atoms in the bond end up having a full octet of electrons.

8. **(C)** As a metal, copper is held together by metallic bonds, which involve delocalized d-orbital electrons.

DETAILED EXPLANATIONS OF ANSWERS

9. **(B)** Nonvascular plants are known as bryophytes (mosses). They lack tissue that will conduct water or food.

10. **(C)** Angiosperms are those plants that produce flowers as reproductive organs.

11. **(C)** Angiosperms (flowering plants) have two main systems—the shoot system, which is mainly above ground and the root system below ground.

12. **(D)** Gymnosperms produce seeds without flowers.

13. **(D)** Gymnosperms produce seeds without flowers, which include conifers (cone-bearers) and cycads.

14. **(D)** Conduction is the transfer of molecules by collisions, passing heat through one material into another.

15. **(B)** Convection is caused by the flow of heated liquid or gas through a volumetric medium.

16. **(A)** Radiation is waves traveling through space to transfer heat away from the energy source.

17. **(B)** The manganese ion in potassium permanganate has a +5 oxidation state, and is therefore readily reduced. When it is reduced, it forces another species to be oxidized. Therefore, the permanganate ion is a very strong oxidizing agent.

18. **(C)** The magnesium atom that is combined with oxygen has a +2 oxidation state. Oxygen carries a –2 oxidation state, and the sum of the oxidation states of oxygen and magnesium must total the charge on the compound, which is 0.

19. **(A)** Hydrogen gas by definition, since it is a standard, has an oxidation potential of 0.

20. **(B)** One of the modes of population growth is represented by the exponential curve (or J-curve). The rate of growth accelerates over time since there are no limiters of growth.

21. **(C)** The rate of increase within a population is represented by the birth rate minus the death rate.

22. **(D)** Mortality is the death rate within a population.

23. **(A)** Natality is the birth rate within a population.

24. **(E)** Another mode of population growth is represented by the logistic curve (or S-curve) for populations that encounter limiting factors in which acceleration occurs up to a point and then slows down.

25. **(D)** The acceleration due to gravity equals approximately 10 meters per second squared. Therefore, in 0.5 seconds (one half second) the ball will have traveled half of 10 meters, or 5 meters.

26. **(D)** The only answer that causes any change in the entire gene pool is mutation (D). A mutation adds or subtracts (or changes) a trait from the existing genes within the pool. Genetic drift, allopatric speciation, and sympatric speciation are all forms of natural selection that cause changes in populations of already expressed traits.

27. **(C)** All of the answers are true except choice (C) because meiosis does not ensure the same kinds of chromosomes; in fact it ensures that there is a variety in the genetic code of different chromosomal material.

28. **(B)** Echinodermata is the class including sea stars and sand dollars that clearly show radial symmetry (like pieces of pie). The rest of the choices show bilateral symmetry where sides are mirror images of the other (like humans).

29. **(B)** Meiosis results in genetic variation since there is a division and sorting of chromosomes resulting in a variety of eggs and sperm (or pollen) that can recombine in new patterns. Mitosis produces identical daughter cells, and metaphase and prophase are parts of that process. Interphase is the growth portion of the cell cycle.

30. **(E)** The evolution of plant species is considered to have begun with aerobic prokaryotic cells.

31. **(A)** A sound wave is a compression or longitudinal wave, which means that it compresses and rarefies as it moves through a medium. Transverse waves oscillate up and down as they move through a medium. The other three terms are not commonly used to label sound waves.

32. **(A)** The desert areas increase evaporation of the water that is present, causing a concentration of salts.

33. **(C)** The property of an object that determines the object's resistance to motion is its mass. A large watermelon has a far greater mass than the pellet of lead shot, the golf ball, the feather, or the sheet of notebook paper. The mass of an object never changes. Its mass is equal to the object's weight divided by the acceleration due to gravity (g) at its current position in space.

DETAILED EXPLANATIONS OF ANSWERS | 197

34. **(C)** The placenta is the connection between the mother and embryo; it is the site of transfer for nutrients, water, and waste between them.

35. **(E)** Gregor Mendel studied the relationships between traits expressed in parents and offspring, and the genes that caused the traits to be expressed.

36. **(B)** The first cells to evolve were most likely unspecialized. Since the Earth's atmosphere was most likely lacking in oxygen, it is presumed that preplant cells were also anaerobic. Early cells were also small, aquatic, and prokaryotic.

37. **(D)** A branch of bipedal primates gave rise to the first true hominids about 4.5 million years ago. The earliest known hominid fossils were found in Africa in the 1970s. The well-known "Lucy" skeleton was named Australopithecus afarensis. It was determined from the skeleton of Australopithecus that it was a biped that walked upright. It had a human-like jaw and teeth, but a skull that resembled that of a small ape. The arms were proportionately longer than humans', indicating the ability to still be motile in trees. The fossilized skull of Homo erectus, the oldest known fossil of the human genus, is thought to be about 1.8 million years old. The skull of Homo erectus was much larger than Australopithecus, about the size of a modern human brain. Homo erectus was thought to walk upright and had facial features more closely resembling men than apes. A third fossil, the oldest to be designated Homo sapiens, is also called Cro-Magnon man, with a brain size and facial features comparable to modern men. Cro-Magnon Homo sapiens are thought to have evolved in Africa and migrated to Europe and Asia approximately 100,000 years ago.

38. **(C)** Acetylene shows a triple bond between the two carbons, which contains two pi bonds. In multiple bonds, the first bond is a sigma bond in which electrons are shared along the internuclear axis. Any additional bonding is created by the sideways overlap of unhybridized *p*-orbitals above and below the internuclear axis. None of the other options for answers contain multiple covalent bonds between any two atoms.

39. **(D)** Equinoxes are described as the two points on the celestial sphere where the ecliptic crosses the celestial equator. Since the crossing occurs on March 21 in this question, it is the spring equinox. Solstices occur in the summer and winter, and correspond to the ecliptic being farthest from the celestial equator. In general, when this crossing occurs on March 21, the Moon is not necessarily aligned with the Earth and the Sun, so the odds of an eclipse occurring are very low. The autumnal equinox occurs at the other end of the celestial sphere, on September 22.

40. **(D)** The DNA of eukaryotes is organized into chromosomes within the nucleus.

41. **(B)** Inorganic cofactors are small non-protein molecules that promote proper enzyme catalysis. These molecules may bind to the active site or to the substrate itself. The most common inorganic cofactors are metallic atoms such as iron, copper, and zinc.

42. **(E)** Environmental conditions such as heat or acidity inhibit enzymatic reactions by changing the shape of the active site and rendering the enzyme ineffective. Certain chemicals inhibit enzymatic reactions by changing the shape of the enzyme's active site. If there is a lack of substrate, the enzyme will have no substance to affect. Thus, a large amount of enzyme is the only factor that will not inhibit enzymatic reactions.

43. **(D)** A ketone group has a double bond between oxygen and a carbon atom that is imbedded within a carbon chain.

44. **(A)** An alcohol group has a hydroxide group (OH) with the second bond from oxygen going to a carbon atom.

45. **(C)** An ester group is an ether with an additional double bond to an oxygen from an adjacent carbon.

46. **(B)** An aldehyde group has a double bond between an oxygen atom and a carbon that is one end of a carbon chain, or not otherwise bonded to any other carbons.

47. **(E)** The process of photosynthesis is the crucial reaction that converts the light energy of the Sun into chemical energy that is usable by living things.

48. **(B)** Entropy is defined as the new state variable that describes the increase of disorder in a system. Stated positively, it describes the amount of increase in the statistical disorder of a physical system.

49. **(C)** The development of photosynthetic cells took place before the development of eukaryotes. The first photosynthetic cells were prokaryotic.

50. **(A)** Energy flows through the entire ecosystem in one direction—from producers to consumers and on to decomposers through the food chain.

51. **(B)** The eclipse must be a solar eclipse because the Moon is between the Earth and the Sun. The solar eclipse is total because the Moon's disk covers the Sun from our view completely. Parallax has to do with the apparent displacement of a celestial object.

DETAILED EXPLANATIONS OF ANSWERS | 199

52. **(C)** The celestial poles are extensions of Earth's north and south geographic poles up into the sky. The celestial equator identifies the projection of Earth's equator. The zenith is the point directly overhead above an observer. The ecliptic is the line that describes the Sun's orbit across the celestial sphere. The zodiac is the name for the annual cycle of twelve stations along the ecliptic that the Sun and planets travel along the celestial sphere.

53. **(D)** Cellular respiration is the process that releases energy for use by the cell. There are several steps involved in cellular respiration; some require oxygen (aerobic) and some do not (anaerobic).

54. **(A)** When tRNA anticodons line up with corresponding mRNA codons, it is the last step in the transcription process before translation begins. Translation begins as a ribosome attaches to the mRNA strand at a particular codon known as the start codon. This codon is only recognized by a particular initiator tRNA. The ribosome continues to add tRNA whose anticodons complement the next codon on the mRNA string. A third type of RNA is utilized at this point, ribosomal RNA or rRNA. Ribosomal RNA exists in concert with enzymes as a ribosome. In order for the tRNA and mRNA to link up, enzymes connected to rRNA at the ribosome must be utilized. Ribosomal enzymes also are responsible for linking the sequential amino acids into a protein chain. As the protein chain is forming, the rRNA moves along the sequence, adding the amino acids that are designated by the codons on the mRNA. At the end of the translation process, a terminating codon stops the synthesis process, and the protein is released.

55. **(A)** The most likely sequence leading to human evolution begins with coacervates, then eukaryotes, plants, fish, amphibians, mammals, primates, and finally, man.

56. **(A)** The hydrogen atoms in water molecules have a partial positive charge, while the oxygen atoms in water have a partial negative charge, causing polarity. This polarity allows the oxygen of one water molecule to attract the hydrogen of another. The partial charges attract other opposite partial charges of other water molecules allowing for weak (hydrogen) bonds between the molecules. *Inert* means non-reactive; it does not explain the attraction between H and O. There are no ionic bonds within water molecules, only covalent bonds. A crystal structure forms in ice because of the attraction of hydrogen bonds; the crystal structure does not cause the attraction. Brownian motion is the random movement of atoms or particles caused by collisions between them; it does not explain the attraction between atoms or molecules.

57. **(A)** Using Ohm's law, V = IR, we solve for I=V/R with V = 12 V and R = 8 ohms. I = 12.0 V/8.0 ohms = 1.5 V/ohm = 1.5 A. The other answers are results of erroneous calculations or misapplications of Ohm's law.

58. **(D)** Earth is approximately 70 percent water. Most of the water is in the oceans. Land makes up the remaining 30 percent.

59. **(A)** Traits, such as height and skin color, are produced from the expression of more than one set of genes and are known as polygenic traits.

60. **(B)** Evolution is driven by the process of natural selection, a feature of population genetics first popularized by Charles Darwin in his book *The Origin of Species* (published in 1859).

61. **(D)** The horn is not a simple machine. Simple machines are mechanical devices that alter the magnitude and direction of a force. They are represented by the following six objects: (1) lever, (2) wheel and axle, (3) pulley, (4) inclined plane, (5) pulley, and (6) screw. On a bicycle, the tire is a type of wheel. The pedal mechanism is a wheel and axle as well as a gear mechanism (two simple machines). The rear wheel gear mechanism is obviously one of the simple machines. The kickstand is a lever.

62. **(C)** Since nitrogen is in the second row, its highest-energy electron is at n = 2, which is the first number. The second number signifies that its outer electron is in a *p*-orbital. The third number indicates the third *p*-orbital to receive an electron, since nitrogen is the third element in the *p*-block in the periodic table. The last number is the magnetic spin quantum number and signifies that the highest energy electron is the only electron in the orbital.

63. **(B)** Although there is a transition from one state of matter to another, there is no term "transition phase" to describe this process. The term fission is used in nuclear physics to describe the breaking apart of nuclear particles. It is not used to describe changes of state of matter. Specific heat is the amount of energy in calories required to raise one gram of a substance by 1°C. Equilibrium is the state of a system or body at rest or lacking acceleration which results when all the forces acting upon it are equal to zero and the sum of all of the torques equals zero. In chemistry, it is the state of a reaction when all the products and reactants are balanced.

64. **(E)** A simple cubic crystal is the only unit cell mentioned that has, on average, one atom per unit cell. A simple cubic unit cell has one atom in each of the eight corners of the unit cell, but only one-eighth of each of those atoms is ascribed to that particular unit cell. A face-centered unit cell

has four atoms per unit cell, while a body-centered unit cell has two atoms per unit cell.

65. **(E)** The alimentary canal is also known as the gastrointestinal (GI) tract and includes the mouth, pharynx, esophagus, stomach, small intestine, and large intestine.

66. **(D)** Proteins are catalysts in the laboratory, so the formation of proteins in the laboratory under conditions presumed to be representative of early Earth history is key evidence of the evolution of life on Earth. Stanley Miller provided support for Oparin's hypotheses in experiments in which he succeeded in producing amino acids by exposing simple inorganic molecules to electrical charges similar to lightning. Sidney Fox conducted experiments that proved ultraviolet light may induce the formation of dipeptides from amino acids. He also showed that polyphosphoric acid could increase the yield of these polymers, a process that simulates the modern role of ATP in protein synthesis. Researcher Cyril Ponnamperuma demonstrated that small amounts of guanine formed from the thermal polymerization of amino acids.

67. **(C)** This is known as incomplete dominance. Neither white nor red is dominant over the other.

68. **(B)** RW is a symbol for genotype. In this case, the RW genotype produces a pink phenotype. R and W represent the alleles for red and white, respectively.

69. **(B)** Both parents have two alleles that are the same; thus they have homozygous genotypes for color.

70. **(D)** While (A), (B), and (C) could have been true, a red snapdragon could have been produced by any of those choices. However, a white snapdragon cannot produce a red snapdragon as an offspring even if paired with a red snapdragon.

71. **(E)** If two of the heterozygous offspring of an incomplete dominant trait are bred, the Punnett square would be

	R	W
R	RR	RW
W	RW	WW

The phenotypic ratio of the offspring, then, is one-fourth red one-half pink, and one-fourth white, a 1:2:1 ratio—1 red: 2 pink: 1 white.

72. **(D)** A habitat refers to the physical place where an organism lives. A species' habitat must include all the factors that will support its life and reproduction.

73. **(A)** Nickel has an unpaired electron in its *d*-orbital, which induces a magnetic field and causes it to be attracted to another magnetic field. Zinc chloride does not have such an unpaired electron and is not attracted to a magnet.

74. **(C)** The heat absorbed by the water equals the mass of the water multiplied by the specific heat of water times the increase in temperature change. The 840 J added to the water is enough heat to raise the temperature of 20 g of water by 10°C.

75. **(B)** A light-year is a distance, not a speed or velocity. It is also not a time measurement. A light-year represents the measurement of the distance to one celestial body from another celestial body, measured as a multiple or fraction of the distance light travels in one year. It is equal to $(3.0 \times 10^8 \text{ m/s})(3.6 \times 10^3 \text{ s/hr})(2.4 \times 10^1 \text{ hr/day})(3.65 \times 10^2 \text{ days/yr})(1 \text{ yr})$ = 95×10^{14} m = 9.5×10^{15} m = 9.5×10^{12} km = 9,500,000,000,000 km or 9.5 trillion kilometers.

76. **(D)** Since the half-life of C 14 is 5,730 years, that means it takes 5,730 years for half the C 14 to decay to C 12. Since the end sample has ¼ C 12, we see that two half-life periods would have occurred. The first would have decayed the sample to half of each type, and the second time period would degrade the half sample to a quarter sample. Two half-life periods is a total of 11,460 years.

77. **(D)** An example of igneous rock is granite. Marble is a metamorphic rock, while limestone, cement, and sandstone are all sedimentary rocks.

78. **(B)** Although it appears that galaxies are moving apart, Hubble's law explains that this apparent motion is the expansion of the space between them, not the motion of the galaxies themselves. Hubble's law states that the redshift in light coming from a distant galaxy is directly and linearly proportional to its distance from Earth.

79. **(D)** There is no epistematic region in plant roots. Roots have four major structural regions that run vertically from bottom to top. The root cap is composed of dead thick-walled cells and covers the tip of the root, protecting it as the root pushes through soil. The meristematic region is just above the root cap. It consists of undifferentiated cells that undergo mitosis, pro-

viding the cells that grow to form the elongation region. In the elongation region, cells differentiate, large vacuoles are formed, and cells grow. As the cells differentiate into various root tissues, they become part of the maturation region.

80. **(B)** Gregor Mendel determined that one gene is sometimes dominant over another gene for the same trait (i.e., it expressed itself over the other). This is known as the law of dominance, Mendel's second law of inheritance.

81. **(B)** Electrons have very small mass, much less than either protons or neutrons. They are found orbiting in a cloud surrounding the nucleus. Electrons are negatively charged. An ion is an atom with a greater or fewer number of electrons than the standard atom for the element, causing a charge of positive or negative due to the unequal number of protons and electrons. An atom's valence number is the number of electrons in its highest (not lowest) energy level.

82. **(B)** Astrophysicists speculate that just 1/100 of a second after the big bang, the entire universe was filled with elementary particles (i.e., electrons, protons, positrons, nutrinos, photons, etc.). At 1 second, the temperature was 10^{10} K, which was still too hot for neutrons and protons to stay together in the nuclei by the strong nuclear force. After about 3 minutes, it had cooled to about 10^9 K. Electrons and positrons were annihilated, generating photons, neutrinos, anti-neutrinos, and a small number of neutrons and protons. The universe was now as hot and dense as the core of a star undergoing nuclear fusion. However, it wasn't until 300,000 years later that the nuclei could hold on to the electrons, clearing the vast fog. The process then continued to cool, coalesce, and expand through fusion, forming stars and galaxies in the visible universe. Fission does not occur until much later in stellar evolution. Spontaneous generation, disproven by Pasteur, has nothing to do with the evolution of stars. Combustion is common in various exothermic processes in a general way. Stellar explosions indicative of the latter stages of stellar evolution generally occur toward the end of a star's life, not at the origination.

83. **(A)** The answer is that entropy increases as a natural process goes from one equilibrium state to another. This is the only direction that heat flow will go in an irreversible process.

84. **(D)** Using Ohm's law, $V = IR$, solve for $R = V/I$ with $V = 120$ V and $I = 10$ A. $R = 120$ V / 10 A $= 12$ V/A or 12 ohms. All of the other answer choices are results of erroneous calculations or misapplications of Ohm's law.

85. **(B)** Stomata are openings in the leaf surface that allow for exchange of water and gases.

86. **(C)** The individual we recognize as an adult fern is actually the mature sporophyte.

87. **(A)** Hydrogen gas is the least polar because it is bonded to itself as a diatomic molecule. Therefore, there is no difference in electronegativity between the two atoms in the bond, and it is entirely nonpolar.

88. **(E)** Both a gas and plasma have no defined shape or volume. A solid has a definite shape and volume. A liquid holds the shape of its container and thus has a definite volume. A plastic is not a state of matter and is confused with the fourth state of matter, plasma.

89. **(B)** Conditions such as temperature, pH, water balance, and sugar levels must be monitored and controlled in order to keep them within the accepted ranges that will not inhibit life. Cells and living organisms have homeostatic mechanisms that serve to keep body conditions within normal ranges.

90. **(B)** The cerebrum controls sensory and motor responses, memory, speech, and intelligence factors. It does not control involuntary muscles.

91. **(C)** Fog consists of a visible collection of minute water droplets suspended in the atmosphere near the Earth's surface. It occurs when atmospheric humidity combines with a warm layer of air that is transported over a cold body of water or land surface.

92. **(E)** Due to the high specific heat capacity of water, land surfaces will heat up more rapidly than water surfaces. Since warmer air is less dense than cooler air, this uneven heating along coastal areas causes vertical expansion of the isobaric field over the land and compression of the field over the water. This forms an elevated area of high pressure over the land and an elevated area of low pressure over the water. The pressure gradient force acts on the air and moves it from higher pressure to lower pressure. The net movement of air toward the low pressure aloft induces an area of high pressure on the water surface, while the net movement of air away from the high pressure aloft induces an area of low pressure on the land surface. The pressure gradient force now begins to move the surface air from the higher pressure on the water surface towards the lower pressure on the land surface. It is this surface flow of air that is the sea (or lake) breeze. The greater the contrast is between water temperature and land temperature, the stron-

ger the breeze. At night, the land will cool faster than the adjacent water, reversing the process and forming a land breeze from the land towards the water.

93. **(D)** The electrical properties of cardiac muscle tissue cause the beating of the heart muscle that results in the pumping of blood through the body.

94. **(C)** Ingested food does not pass through the salivary glands. The saliva secreted from these glands enters the digestive tract and helps digest the food.

95. **(C)** Carrying capacity (designated by K) is the number of organisms that can be supported within a particular ecosystem.

96. **(D)** Quarks, bosons, neutrinos, and positrons are all subatomic particles. An electron is an atomic particle that represents the negatively charged part of an atom.

97. **(A)** Digestive enzymes are released by the pancreas and gall bladder into the small intestine.

98. **(A)** Insects have spiracles that allow for gas exchange.

99. **(A)** Over 70 percent of Earth's surface is covered by water. The great majority of Earth's photosynthesis takes place near the surface of the oceans, with 70 percent of the entire amount of it taking place in the euphoric zone to a depth of 100 m. Many people mistakenly think that the tropical rainforests and deciduous forests are responsible for most of the production of oxygen during photosynthesis, but this is not so. The tundra and deserts are even more obviously incorrect because the amount of plant life in these biomes is very sparse.

100. **(C)** A replacement reaction occurs when a compound is broken down into its components and recombined with another reactant—as shown by the equation AB 1 C → AC 1 B. The reaction A 1 B → AB represents a combination reaction; AB → A 1 B and AB 1 CD → A 1 B 1 C 1 D are both decomposition reactions; and A 1 C → AC 1 B is not a possible reaction since it is not with the first law of thermodynamics, which states that matter cannot be created or destroyed in a chemical reaction. Since B was not a reactant, it cannot be a product.

101. **(D)** Ecology is literally "the study of" (ology) a "place to live" (eco).

102. **(E)** Each species is a distinct group of individuals that are able to mate and produce viable offspring.

103. **(B)** Einstein's energy equation, $E = mc^2$, is based on the premise that the speed of light for all observers is constant, and that energy is directly proportional to mass. As energy increases, so does its mass. The speed of light is considered constant; the mass and energy values must increase or decrease directly and proportionately.

104. **(D)** Plants that produce their own food through photosynthesis are known as autotrophs. (Mushrooms are fungi; they do not produce their own food.)

105. **(A)** The speed of light is the same in all directions. In a vacuum, the speed of light remains constant. The speed of light in a liquid is slower than in air. The speed of light is the same on the Moon as it is on the Earth.

106. **(B)** The principle of supererogation is completely fictitious, so this is the correct answer. The principle of original horizontality states that if rock layers are not laid horizontally, then something forced the structures to move. The principle of lateral continuity means that sediments are originally deposited in layers that extend laterally in all directions and eventually thin out. The principle of fossil succession says that if rocks are undisturbed, the oldest layers of rock should be on the bottom and will have older fossils in them. The principle of uniformitarianism states that present geological events explain former ones. It is popularly stated as the principle that "the present is the key to the past."

107. **(A)** A redshift in the electromagnetic spectrum of the light from a faraway star means that the star is moving away from the observer. In the 1920s, Edwin Hubble (1889–1953) observed that galaxies around the Milky Way were moving away from us because of this redshift, and those farther away are moving away from us even more rapidly; it is not moving toward us. The redshift does not indicate that a star is ready to explode, is a dwarf star, or is a giant nova.

108. **(C)** These eras are parts of the Phanerozoic Eon and represent the past 570 million years. Cenozoic represents "recent life," Mesozoic represents "middle life," and Paleozoic represents "old life." There were complex life-forms during all three of these eras.

109. **(A)** Parasitism is symbiosis in which one species benefits, but the other is harmed.

110. **(B)** An axis is defined as a line connecting two poles. Earth has an axis connecting the North Pole and the South Pole. The other choices are incorrect.

111. **(D)** There are estimated to be about 100 billion stars in the universe. To count to a billion, if a person counted one number per second it would take over 34 years. To then count to 100 billion would take $34 \times 100 = 3,400$ years! In other words, a billion is a huge number.

112. **(B)** The astronaut will feel as if he is being pushed forward toward the front of the spacecraft. The others are incorrect directions of motion. There is a negative acceleration (deceleration) and thus a negative force occurring since the vehicle is slowing down.

113. **(C)** Blue-green algae is a prokaryotic organism in the Kingdom Monera. Prokaryotes have no nucleus-or-membrane-bound organelles.

114. **(B)** The celestial sphere is an imaginary sphere centered on and surrounding Earth upon which the background stars are projected. The Sun, Moon, planets, and other celestial bodies seem to move relative to this background of fixed stars. It is a model used to describe positions and motions of astronomical bodies. A geodesic dome is not an astronomical tool, but is simply an architectural structure. A stellar ball is a made-up term. The Cartesian coordinate system is a mathematical system used to plot points on an x-y graph. The equatorial coordinate system is another mathematical system used to plot specific celestial bodies in the sky using time and angular coordinates of right ascension and declination. It relates the stars to their apparent motion in the sky along the celestial sphere relative to the celestial equator. It is used to make measurements within the celestial sphere, but is not actually the sphere itself.

115. **(B)** The "greenhouse effect" is the result of radiation being blocked from leaving the atmosphere because of the ozone layer. Depletion of the ozone layer would impact the "greenhouse effect," not cause it. The "greenhouse effect" is also caused mainly by other atmospheric gases such as water vapor and carbon dioxide. This depletion will increase radiation because the hole in the ozone layer will get larger. Acid rain is primarily caused by oxides of sulfur and nitrogen due to pollution. Photochemical reactions involving oxides of nitrogen and chloro-fluorocarbons play a major role in ozone destruction. These chemicals react with ozone and lead to a net decrease in ozone concentration in the stratosphere. Since ozone is a strong absorber

of incoming ultraviolet (UV) radiation, any decrease in ozone concentration will increase the amount of UV radiation reaching the surface. This will significantly increase the risk of skin cancer from exposure to the Sun. With this in mind it is easy so see why choice (D) is incorrect. However, the ozone depletion has no impact on global warming directly.

116. **(C)** The pressure gradient force plays no significant role in vertical air circulations. Since temperature decreases with height, as the valley air warms during the day, it becomes less dense than the air along the adjacent hillsides and begins to flow up the surrounding slopes. At night, air in the hills cools faster than the air in the valleys. Gravity pulls the cooler, denser air down the hillsides into the valleys. It is low pressure near the top of the mountain that forces breezes up from the valley, so choice (E) is wrong.

117. **(A)** Only the size of the atom increases as you move down and to the left in the periodic table.

118. **(C)** The greater the mass of an object, the greater the force necessary to change its state of motion. Newton's first law of motion states that an object in motion will stay in motion and an object at rest will stay at rest until acted upon by an outside force. Since Newton's second law of motion states that Force = (mass)(acceleration), then in order to change its motion, as the object increases in mass, more force will be necessary to alter its acceleration. Its position may be changed slightly with the same amount of force as might be applied to a smaller object, so changing its position is not the correct answer. Greater force may not change its shape at all, and will not change its density.

119. **(A)** The top of the sine wave is the crest. The bottom is the trough. The amplitude represents the height of the wave. The period is the wave's cycle length. The frequency is the number of cycles per second.

120. **(D)** The water molecules, which are the medium, are not carried but the wave peaks do move toward shore. Answer choice (A) erroneously states the opposite of what is mentioned in the question. Although the ocean is somewhat mixed each day, this mixing is due to turbulence and currents, not waves. Similarly, debris washes ashore by turbulence and currents and not wave action. Waves also do not move swimmers, nor water; the waves travel through the water and the swimmer will bob up and down but will not be carried by the wave. Again, currents and turbulence are responsible for moving both the water and swimmers.

PRACTICE TEST 2

CLEP Natural Sciences

Also available at the REA Study Center (*www.rea.com/studycenter*)

This practice test is also available online at the REA Study Center. The CLEP Natural Sciences test is only offered as a computer-based exam; therefore, we recommend that you take the online version of the practice test to receive these added benefits:

- **Timed testing conditions** – Gauge how much time you can spend on each question.
- **Automatic scoring** – Find out how you did on the test, instantly.
- **On-screen detailed explanations of answers** – Learn not just the correct answer, but also why the other answer choices are incorrect.
- **Diagnostic score reports** – Pinpoint where you're strongest and where you need to focus your study.

PRACTICE TEST 2

CLEP Natural Sciences

(Answer sheets appear in the back of the book.)

TIME: 90 Minutes
        120 Questions

DIRECTIONS: Each of the following groups of questions consists of five lettered terms followed by a list of numbered phrases or sentences. For each numbered phrase or sentence, select the one choice that is most clearly related to it. Each choice may be used once, more than once, or not at all. Fill in the corresponding oval on the answer sheet.

Questions 1–5

    (A) Primary oocytes
    (B) Cleavage
    (C) Secondary spermatocytes
    (D) Oogenesis
    (E) Polar body

1. _____ Present in reproductive organs at birth

2. _____ Division of the zygote into multiple cells

3. _____ Haploid cells that will develop into male gametes

4. _____ Infertile cell resulting from meiosis II in females

5. _____ Develop into spermatid

Questions 6–9

(A) Alpha decay
(B) Beta (β-) decay
(C) Positron decay
(D) Neutron capture
(E) Gamma radiation

6. _____ The nuclear reaction that turns a proton into a neutron

7. _____ The emission of very high energy photons

8. _____ The nuclear reaction that gives off the nucleus of a helium atom

9. _____ The nuclear reaction that turns a neutron into a proton

Questions 10–13

(A) $Ag^+ + I^- \rightarrow AgI$
(B) $HC_2H_3O_2 + OH^- \rightarrow C_2H_3O_2^- + H_2O$
(C) $NaCl(l) \rightarrow Na(l) + Cl_2(g)$
(D) $Mg + O_2 \rightarrow MgO$
(E) $Fe_2^+ + 8H^+ + MnO_2^- \rightarrow Fe^{3+} + Mn^{2+} + 4H_2O$

10. _____ Two electrons are transferred in this reaction.

11. _____ This reaction results in a basic solution.

12. _____ This reaction yields a cloudy product.

13. _____ This reaction takes place in an electrolytic cell with the input of electrical energy.

Questions 14–18

(A) Limiting factors
(B) Law of minimums
(C) Homeostasis
(D) Density
(E) Law of tolerance

14. _____ Biotic and abiotic influences that cause a disturbance in an ecosystem that affects the population growth rate

15. _____ Dynamic balance achieved within an ecosystem functioning at its optimum level

16. _____ The tendency for the resource in shortest supply to limit the population growth

17. _____ Population growth is limited by having more of a substance than is required

18. _____ Number of organisms per area

DIRECTIONS: Each of the questions or incomplete statements below is followed by five possible answers or completions. Select the best choice in each case and fill in the corresponding oval on the answer sheet.

19. Which of the following is NOT an example of homologous structures?

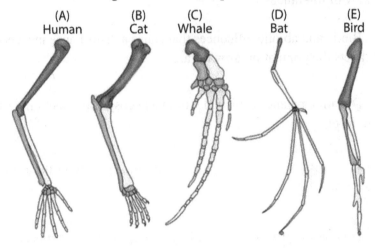

(A) Human (B) Cat (C) Whale (D) Bat (E) Bird

20. The mass of a ball rolled on a potentially frictionless horizontal surface is 9.1 kg while its acceleration is 3.2 m/s². What is the force of the rolling ball?

(A) 29.1 N
(B) 2.8 N
(C) 12.3 N
(D) 5.6 N
(E) 9.6 N

21. Which of the following Earth "spheres" contains molten magma?

(A) Troposhere
(B) Lithosphere
(C) Biosphere
(D) Hydrosphere
(E) Stratosphere

22. Which of the following shows logistic population growth?

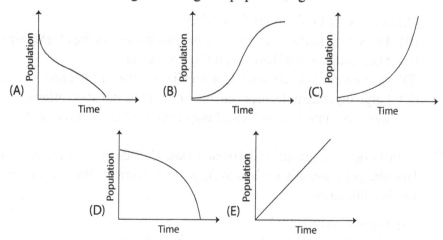

23. How much energy is consumed by each level in the energy pyramid?

   (A) 10%
   (B) 90%
   (C) 00%
   (D) 50%
   (E) 0%

24. Which of the following is within the phylum Chordata?

   (A) Crabs
   (B) Flat worms
   (C) Nematodes
   (D) Mollusks
   (E) Perch

25. Which of the following organs does NOT function as an organ of the immune system to defend the body from infection?

   (A) Tonsils
   (B) Lymph nodes
   (C) Spinal cord
   (D) Spleen
   (E) Thymus

26. Which of the following is NOT included in Oparin's hypothesis?

    (A) H_2O existed only in the form of ice.
    (B) The young Earth had very little oxygen present in the atmosphere.
    (C) The Earth is more than four billion years old.
    (D) Heat energy was abundantly available because of the Earth's cooling.
    (E) Large organic molecules capable of dividing and absorption (coacervates) became plentiful, resulting in eventual evolution of early life.

27. Both London, England, and Hudson Bay, Canada, are on the same north latitude, yet London's winters average 35°F warmer. What could account for this difference?

    (A) England is an island.
    (B) The waters of the Gulf Stream wash the English coast.
    (C) Latitude is not a factor in determining temperature.
    (D) Warm winds from France moderate the weather in England.
    (E) The waters of the Gulf Stream wash the Canadian coast.

28. Which of the following is a true statement regarding fundamental atomic particles?

    (A) There are five different types of quarks.
    (B) Each quark is called *a flavor*.
    (C) The five flavors of quarks are *up, down, strange, ceiling,* and *floor*.
    (D) The basic combination structure of three quarks for a proton is one *up* quark, one *down* quark, and one *strange* quark.
    (E) The structure of the atom is nearly solved since quarks are the most fundamental property that exists.

29. Which of the following best explains why solid copper is a strong conductor, but solid copper sulfate is NOT?

    (A) The sulfate blocks the flow of electrons.
    (B) The unpaired electrons in copper's outer shell allow electricity to flow through the atom.
    (C) The d-level electrons in metallic bonds are delocalized.
    (D) Any compound with sulfur would not conduct electricity.
    (E) The copper is more diluted in copper sulfate.

30. Members of which of the following categories are most closely related?

    (A) Phylum
    (B) Genus
    (C) Kingdom
    (D) Class
    (E) Order

31. After a forest fire, a meadow community develops and is later replaced by a temperate forest community. This process is called

    (A) commensalism
    (B) succession
    (C) dynamic equilibrium
    (D) climaxing
    (E) mutualism

32. If the nucleus of an oxygen isotope (with 8 protons and 8 neutrons) was split exactly in two, what would the new element(s) consist of (neglect consideration of any energy that would be involved)?

    (A) Two oxygen atoms
    (B) Two beryllium atoms
    (C) Two helium atoms
    (D) Two neon atoms
    (E) One oxygen ion

33. Given the order of magnitude of the four fundamental forces, what is their order from weakest to strongest?

    (A) Gravity, strong nuclear, weak nuclear, electromagnetic
    (B) Gravity, weak nuclear, electromagnetic, strong nuclear
    (C) Strong nuclear, electromagnetic, weak nuclear, gravity
    (D) Strong nuclear, weak nuclear, gravity, electromagnetic
    (E) Electromagnetic, gravity, strong nuclear, weak nuclear

34. Which of the following kingdoms contains photosynthetic organisms?

    (A) Monera
    (B) Protista
    (C) Animalia
    (D) Fungi
    (E) Chordata

35. A cup of water is placed in a microwave oven. The timer is set for four minutes, which is plenty of time for it to come to a boil. A second cup is then placed in the same microwave oven containing a piece of a chocolate bar. It melts in less than a minute. The difference in the amount of time required to melt the chocolate versus boiling the water can be BEST explained by which physical characteristic?

    (A) ΔT
    (B) British thermal unit (Btu)
    (C) Specific heat
    (D) Kelvin temperature
    (E) Temperature inversion

36. Which layer of the atmosphere is closest to the surface of Earth?

    (A) Exosphere
    (B) Thermosphere
    (C) Mesosphere
    (D) Stratosphere
    (E) Troposphere

37. Thermodynamics deals with the study of which of the following systems that can be observed and measured in experiments?

    (A) Large-scale response
    (B) Small-scale response
    (C) Small-scale and large-scale responses
    (D) Flow between small-scale and large-scale responses
    (E) Large-scale to small-scale responses

38. Which of the following cell organelles is known as the cell's "powerhouse" because it produces energy for the cell's use?

    (A) Nucleus
    (B) Mitochondrion
    (C) Smooth endoplasmic reticulum
    (D) Ribosome
    (E) Cell membrane

39. Which of the following factors exerts the most influence over limiting cell size?

    (A) A rigid cell wall
    (B) The ratio of surface area to volume of cytoplasm
    (C) Replication of mitochondria
    (D) The chemical composition of the cytoplasm
    (E) The chemical composition of the cell membrane

40. The cause of Earth's magnetic field is

    (A) enormous currents
    (B) space satellites
    (C) still a mystery from deep in Earth's core
    (D) found in the Earth's crust
    (E) huge amounts of iron, nickel, and cobalt

41. Which of the following explains why a sound wave has a higher pitch than the original source?

    (A) The source is standing still.
    (B) The source is moving forward.
    (C) The source is moving away.
    (D) The source is going up and down.
    (E) The source has a higher amplitude than the resultant wave.

42. When water pressure is equal inside and outside the cell, it is said to be

    (A) hydrostatic
    (B) diffuse
    (C) isohydric
    (D) isotonic
    (E) hydrophobic

43. When astronauts sleep aboard the space shuttle, they strap themselves to a wall or a bunk. If one of the thruster rockets were fired, resulting in a change of shuttle velocity and acceleration, any unstrapped sleeping astronaut would be injured by slamming into a shuttle wall. The sleeping astronauts would need to be strapped down because

    (A) their bodies have inertia
    (B) gravity is not strong enough to keep them from touching their beds
    (C) the body functions need to be constantly monitored
    (D) astronauts cannot sleep while floating in free space
    (E) their bodies need a reaction force

44. Which of the following organelles is found in plants only and helps provide rigidity to cells?

    (A) Microtubules
    (B) Cell walls
    (C) Microfilaments
    (D) Centrioles
    (E) Mitochondria

45. Which of the following statements about enzymes is NOT true?

    (A) High temperatures destroy most enzymes.
    (B) Enzymes only function within living things.
    (C) An enzyme is unaffected by the reactions it catalyzes, so it can be used over and over again.
    (D) Enzymes are usually very specific to certain reactions.
    (E) Some enzymes contain a non-protein component that is essential to their function.

46. Which of the following pairs of matter and antimatter are related correctly?

    (A) Proton . . . antiproton
    (B) Electron . . . antielectron
    (C) Neutron . . . neutrino
    (D) Positron . . . negitron
    (E) Neutrino . . . quark

47. The astronomical unit or AU is

    (A) the radius of the Moon
    (B) the average distance of Earth from the Sun
    (C) the force that the Sun exerts on a planet
    (D) a recently discovered satellite of Jupiter
    (E) a quantum measurement of the energy released by a falling asteroid

48. Scurvy is a disease caused by a lack of vitamin C, in which the body is unable to build enough collagen (a major component of connective tissue). The most plausible explanation for this malfunction is

    (A) Vitamin C is an amino acid component of collagen
    (B) Vitamin C is a coenzyme required in the synthesis of collagen
    (C) Vitamin C destroys collagen
    (D) Vitamin C is produced by collagen
    (E) Vitamin C is not related to collagen

49. An ion that attaches to an enzyme to become better able to catalyze a reaction is known as

    (A) a protein
    (B) an inorganic cofactor
    (C) a coenzyme
    (D) a prosthetic group
    (E) a vitamin

50. When a stream full of sediments flows out of a narrow mountain canyon and widens into a stream channel, losing velocity and forming large fan-shaped deposits, it is called a(n)

    (A) delta
    (B) tributary
    (C) alluvial fan
    (D) topset bed
    (E) meandering stream

51. A thermodynamic process that does not involve heat transfer is called a(n)

   (A) equilibrium process
   (B) net-sum process
   (C) adiabatic process
   (D) cooled-state process
   (E) steady-state process

52. Monotremes are mammals that lay eggs. The feature of egg laying in this case most likely suggests

   (A) a common ancestry with birds
   (B) a convergent evolution of analogous traits
   (C) a common ancestry with reptiles
   (D) an evolution of homologous traits
   (E) a mutational trait

53. Which of the following gases would be expected to show the greatest deviation from ideal behavior?

   (A) H_2
   (B) H_2S
   (C) H_2O
   (D) O_2
   (E) CH_4

54. If you were aboard a spacecraft that is traveling away from Earth at 9.8 m/s², you would feel the same force as if you were standing on Earth's surface. This is a representation of

   (A) Einstein's equivalence principle
   (B) a total inaccuracy
   (C) the results of the Michelson-Morley experiment
   (D) the relationships associated with light, mass, and energy in Einstein's general theory of relativity
   (E) the Doppler effect

55. The synthesis of ATP molecules to store energy is an example of

    (A) anabolism
    (B) catabolism
    (C) enzyme response
    (D) lysis
    (E) inhibition

56. Energy transformations that occur as chemicals are broken apart or synthesized within the cell are collectively known as

    (A) catabolism
    (B) anabolism
    (C) metabolism
    (D) synthesis
    (E) lysis

57. Archimedes' principle says that an object is buoyed up by a force that is equal to the

    (A) weight of the fluid displaced
    (B) volume of the fluid displaced
    (C) mass of the fluid displaced
    (D) mass of the object
    (E) inverse squared of the density of the fluid

58. When determining the molar volume of a gas by collecting the gas over water, which of the following laws must be considered to take into account the pressure exerted by water vapor in the collection bottle?

    (A) Boyle's law
    (B) Charles's law
    (C) Dalton's law
    (D) Henry's law
    (E) Pascal's law

59. Which of the following is NOT true of an electromagnet?

    (A) When electricity flows through a wire, the wire becomes a magnet.
    (B) An electromagnet is a permanent magnet.
    (C) The strength of an electromagnet is limited and cannot be increased easily.
    (D) Both (B) and (C)
    (E) (A), (B), and (C)

60. Two-thirds of all known active volcanoes are in the

    (A) Atlantic Ocean
    (B) Pacific Ocean
    (C) Gulf of Mexico
    (D) Baltic Sea
    (E) Indian Ocean

61. Which of the following types of waves are longitudinal?

    (A) The direct wave and the reflected wave from a radio station
    (B) P-waves from an earthquake
    (C) Electromagnetic (EM) waves in a vacuum
    (D) Gravitation waves on a dwarf star
    (E) Ocean waves

62. A solid object with a specific gravity greater than 1 will typically do which of the following, based on this parameter alone?

    (A) Sink in liquid water
    (B) Float in liquid water
    (C) Completely and immediately dissolve in liquid water
    (D) Increase the temperature of the water
    (E) Mix evenly and stay mixed in liquid water

63. The continuous cycle of the alternating rise and fall of the sea level, observed along the coastlines and bodies of water connected to the sea, is called

   (A) the tide
   (B) moon exercises
   (C) sun exercises
   (D) a tidal wave
   (E) a tsunami

Questions 64–68

In this illustration "T" stands for tall and "t" for short.

	T	T
t	Tt	Tt
t	Tt	Tt

64. The illustration above is called

   (A) a Mendelian diagram
   (B) a genotype
   (C) a phenotype
   (D) a phenogram
   (E) a Punnett square

65. What is the genotype of each of the parents?

   (A) TT and tt
   (B) Tt
   (C) TT
   (D) tt
   (E) Tt and TT

66. The gametes for this cross will have which possible genes?

   (A) T, T
   (B) T, t
   (C) TT, tt
   (D) t, t
   (E) XY

67. What will be the phenotypic ratios of the offspring in this cross?

   (A) 4 tall: 0 short
   (B) 2 tall: 2 short
   (C) 4 Tt: 0 tt
   (D) 3 Tt: 1 tt
   (E) XY

68. If two of the offspring are crossed, what will the phenotypic ratio of the next generation be?

   (A) 4 Tt: 0 tt
   (B) 1 tt: 2 Tt; 1 TT
   (C) 4 tall: 0 short
   (D) 3 tall: 1 short
   (E) XY

69. Suppose you are in a spacecraft that is flying through space at a speed that makes the clocks on board seem, as seen from Earth, like they are going half as fast from their point of reference. You measure a friend's mass on Earth and your own before you leave as 70 kg. Assuming you did not gain or lose any mass, what will your mass be on board the spacecraft?

   (A) 35 kg
   (B) 70 kg
   (C) 140 kg
   (D) 210 kg
   (E) 420 kg

70. The parallax of a star is

    (A) its temperature divided by its mass
    (B) its mass divided by its temperature
    (C) the angle through which the star appears to move in the course of half a year
    (D) the angle through which the star appears to move in the course of a year
    (E) the luminosity of the star in angstrom units

71. Which kingdom has members that were most likely to have evolved earlier than members of any other kingdom?

    (A) Kingdom Protista
    (B) Kingdom Plantae
    (C) Kingdom Monera
    (D) Kingdom Animalia
    (E) Kingdom Fungi

72. In mechanics, there are three fundamental quantities that can be measured. All mechanical measurements can be reduced to a combination of these three quantities. What are they?

    (A) Weights, distances, and motions
    (B) Solids, liquids, and gases
    (C) Heights, lengths, and widths
    (D) Length, mass, and time
    (E) Space, time, and matter

73. Which of the following demonstrates the highest melting point?

    (A) Hydrogen gas
    (B) Butane
    (C) KI
    (D) SiC
    (E) H_2O

Questions 74–78

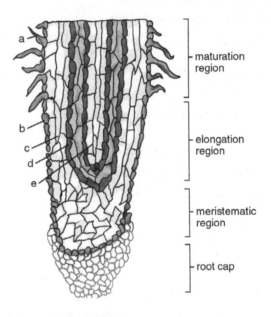

Root Cross-section

Match the tissue type (all are types of primary root tissue) labeled a–e on the diagram (and corresponding to answer choices A–E), with the correct name given in the following five numbered questions.

(A) a
(B) b
(C) c
(D) d
(E) e

74. _____ Vascular cylinder

75. _____ Epidermis

76. _____ Root hairs

77. _____ Endodermis

78. _____ Cortex

79. Astronomers call the visible surface of the Sun the

    (A) photosphere
    (B) plasma
    (C) sunspot
    (D) flare
    (E) corona

80. The weight of an object depends on the

    (A) size of the object and its volume
    (B) mass of an object and its density
    (C) mass of an object and the gravitational force pulling on it
    (D) density of an object and the amount of force it exerts upward against gravity
    (E) type of material in the object and how much pressure is exerted on it

81. A bilayer of phospholipids with protein globules interspersed is characteristic of which of the following organelles?

    (A) Cell (or plasma) membrane
    (B) Mitochondria
    (C) Lysosome
    (D) Chromatin
    (E) Ribosome

82. As light from a star moves outward into space, its energy covers larger and larger areas. The brightness of the star decreases proportionately to the square of its distance from the source. This means that if the distance is quadrupled, the brightness will decrease by a factor of

    (A) 2
    (B) 4
    (C) 8
    (D) 16
    (E) 32

83. The iron-containing molecule that carries oxygen within red blood cells throughout the body via the circulatory system is the

   (A) lymphocyte
   (B) erythrocyte
   (C) gamma globulin
   (D) hemoglobin
   (E) carbohydrate

84. An unknown plant found in the forest has five petals on its flower and a tap root system. Which of the following is most likely true?

   (A) The stem has random arrangements of vascular bundles.
   (B) The plant's seed has only one cotyledon.
   (C) The leaves of the plant have parallel veins.
   (D) The leaves of the plant have networked veins.
   (E) The flowers have no seeds.

85. Which of the following is a characteristic of light?

   (A) Light behaves as a particle.
   (B) Light behaves as a wave.
   (C) Light speeds up and slows down as it moves through a medium.
   (D) Light moves in a medium called luminiferous ether.
   (E) Both (A) and (B) are correct.

86. Why do we see lightning before we hear thunder?

   (A) The lightning occurs closer than the thunder.
   (B) Lightning is composed of multiple strokes.
   (C) Thunder occurs higher in the clouds.
   (D) Because of the differences between the speeds of light and sound.
   (E) Because the sound is slowed down by the liquid drops during the rainstorm associated with the event

87. Temporary movement of a species from one range to another, then back to the original range, is known as

   (A) immigration
   (B) migration
   (C) emigration
   (D) dispersion
   (E) removal

88. Given that a substance has a mass of 6.00 grams, an energy source raises the temperature by 2.345°C and transfers 12 calories of energy to it. If the substance does not melt, thaw, freeze, boil, or condense, what is its specific heat?

    (A) 2.0 cal/g/°C
    (B) .852 cal/g/°C
    (C) 1.17 cal/g/°C
    (D) 2.56 cal/g/°C
    (E) .39 cal/g/°C

89. You have just experienced weightless flight on a zero-gravity plane, which travels in parabolic motion at approximately 600 miles per hour over the Atlantic Ocean. As the plane dived toward the Earth, you floated in the plane's fuselage. Inside the aircraft, you experienced the same absence of gravity that an astronaut does who is flying in the space shuttle as it orbits Earth. This is a representation of

    (A) Einstein's equivalence principle
    (B) a total inaccuracy
    (C) the results of the Michelson-Morley experiment
    (D) the relationships associated with light, mass, and energy in Einstein's general theory of relativity
    (E) the relationships associated with velocity due to gravity in Einstein's special theory of relativity

90. How is the relative strength of a thunderstorm updraft estimated?

    (A) By the intensity of the lightning
    (B) By the height of the cloud base
    (C) By the speed of the wind gusts measured at the surface
    (D) By the size of the precipitation particles (raindrops or hailstones)
    (E) By the speed of the rain droplets as they fall to the surface

91. Plants reproduce through alternation in generations. The diploid generation that produces male and female gametophytes is known as the

    (A) zygotic generation
    (B) sporophyte generation
    (C) gametophyte generation
    (D) second generation
    (E) first generation

92. Suppose that an isotope of carbon has 6 electrons and 6 neutrons. What is its approximate atomic mass?

    (A) 6
    (B) 12
    (C) 18
    (D) 24
    (E) 36

93. Which of the following demonstrates the greatest increase between the atom's second and third ionization energies?

    (A) He
    (B) Na
    (C) Mg
    (D) Al
    (E) N

94. A particular plant has individuals that are either male or female. A male individual of this plant may have all of the following EXCEPT

    (A) a filament
    (B) an anther
    (C) a stigma
    (D) pollen grains
    (E) tube nuclei

95. Since galaxies, planets and stars are bound together by gravity, we don't expand locally even though the universe does. This is because gravity is reduced as the distances are squared. Since this is so, which of the following best explains what the universe is expanding into?

    (A) No one knows because it is too big to see its boundaries
    (B) It appears that the farther we look out into the universe, the more "stuff" we see. There is no end to the matter in the universe.
    (C) Most cosmologists are becoming convinced that the universe has a limit akin to that of a big sphere. It appears to be infinite because of the curve of space.
    (D) Nothing. Since the universe is infinite, nothing exists outside of it. As it expands, it is still infinite, just bigger.
    (E) The universe is in danger of expanding into a black hole that will eventually pull the entire universe into it.

96. Which of the following structural layers of the Earth is liquid?

    (A) crust
    (B) mantle
    (C) core
    (D) lithosphere
    (E) asthenosphere

97. The law of segregation states that

    (A) one gene is usually dominant over the other (expresses itself over the other)
    (B) genes are inherited via a process of multiple alleles
    (C) genes are separated in gamete formation and randomly brought together in fertilization
    (D) gamete formation causes mutation in genetic material of both parents
    (E) mutation of genes is inevitable

98. What is the Celsius equivalent of 54.5°F?

    (A) 22.5°C
    (B) 12.5°C
    (C) 66.1°C
    (D) 62.3°C
    (E) 45.0°C

99. Two carbon rods are put into a solution of $CuSO_4$. The electrodes are connected to a power source, which is turned on for 10 minutes. Which of the following will occur?

    (A) The mass of the cathode will increase.
    (B) The mass of the anode will increase.
    (C) Neither electrode will change in mass.
    (D) A gas will be produced at the cathode.
    (E) No gas will be produced at either electrode.

100. Which of the following are not involved in the immune system?

   (A) Antibodies
   (B) Stem cells
   (C) T cells
   (D) Epithelial cells
   (E) B cells

101. Diamondback rattlesnakes fill a niche in a chaparral biome. Which of the following characteristics does NOT directly contribute to the viability of the snake in its niche?

   (A) Pond water temperature
   (B) Population of hawks
   (C) Population of mice
   (D) Average daily temperature
   (E) Average yearly rainfall

102. To be considered a mineral, the substance must never have been a part of a living organism and must be found in

   (A) rock
   (B) ore
   (C) nature
   (D) sand
   (E) water

103. Which of the following is NOT a trait of an animal?

   (A) Animal cells have organized nuclei and membrane-bound organelles.
   (B) Animal cells do not have cell walls or plastids.
   (C) Animals are only capable of asexual reproduction.
   (D) Animals develop from embryonic stages.
   (E) Animals are heterotrophic (they do not produce their own food).

104. Plants and animals obtain usable nitrogen through the action of

   (A) respiration
   (B) nitrogen fixation by bacteria and lightning
   (C) nitrogen processing in the atmosphere
   (D) digestion
   (E) photosynthesis

105. Why does a cement floor feel cooler to the touch than a wooden floor?

    (A) Because the cement floor never absorbs enough heat to reach room temperature.
    (B) Because the heat conductivity of cement is higher than wood, heat flows more readily through your fingers.
    (C) Because the heat conductivity of cement is higher than wood, heat flows more readily into your fingers.
    (D) Because the heat conductivity of wood is higher than cement, heat flows more readily from your fingers.
    (E) Because the heat conductivity of wood is higher than cement, heat flows more readily into your fingers.

106. As you move across a row in the periodic table, which of the following traits does NOT increase?

    (A) Atomic radius
    (B) Electronegativity
    (C) Electron affinity
    (D) Ionization energy
    (E) Acidity of the oxides

107. If Star X appears to be 2.5 times as bright as Star Y, then the two stars differ in magnitude by

    (A) 0
    (B) 1
    (C) 2.5
    (D) 0.5
    (E) 5

108. Which of the following is NOT a step in the carbon cycle?

    (A) Carbon is taken in by plants and used to form carbohydrates through photosynthesis.
    (B) Carbon is dissolved out of the air into ocean water, combined with calcium to form calcium carbonate, and used by mollusks to form their shells.
    (C) Carbon is taken in by animal respiration and used to form carbohydrates.
    (D) Detritus feeders return carbon to elemental form.
    (E) Burning fossil fuels releases carbon dioxide into the atmosphere where it can be used by plants.

109. The stomach secretes all of the following EXCEPT

   (A) digestive enzymes
   (B) hydrochloric acid
   (C) gastric juices
   (D) acetic acid
   (E) mucous

110. Minerals are identified and classified by all of the following measurable physical properties EXCEPT

   (A) firmness
   (B) cleavage
   (C) fracture
   (D) luster
   (E) color

111. The organelle of a cell that engages in both passive and active transport is the

   (A) rough endoplasmic reticulum
   (B) smooth endoplasmic reticulum
   (C) Golgi complex
   (D) cell wall
   (E) cell (plasma) membrane

112. The path of the Sun across the celestial sphere is called the

   (A) ecliptic
   (B) celestial equator
   (C) equinox
   (D) solstice
   (E) orbital path

113. Which of the following statements is true?

    (A) A niche only includes the physical features of an organism's habitat.
    (B) An organism's habitat must include all the biotic and abiotic factors that will support its life and reproduction.
    (C) A niche refers to the biotic features of an organism's habitat.
    (D) An organism's habitat includes only biotic factors such as weather, temperature, etc.
    (E) An organism's niche does not include its habitat.

114. Compare the bond angles in CH_4 versus XeF_4.

    (A) Both have the same bond angles.
    (B) CH_4 demonstrates 109.5°; XeF_4 demonstrates 90°.
    (C) CH_4 demonstrates 109.5°; XeF_4 demonstrates 120°.
    (D) CH_4 demonstrates 90°; XeF_4 demonstrates 120°.
    (E) CH_4 demonstrates 90°; XeF_4 demonstrates 109.5°.

115. It takes a certain amount of energy to change a sample of liquid to its gaseous state, if it is assumed that the substance can exist in both states. The quantity varies for different substances and is called the

    (A) heat of fusion
    (B) heat index
    (C) heat wave constant
    (D) heat of vaporization
    (E) specific heat

116. A pond ecosystem has sharp boundaries at the shorelines. The sharp boundary of an ecosystem is known as a(n)

    (A) closed community
    (B) open community
    (C) ecotone
    (D) borderline
    (E) system

117. The fossil of a fish skeleton is found in a limestone bed. The fish most likely belonged to the class

    (A) Chondrichthyes
    (B) Reptilia
    (C) Osteichthyes
    (D) Aves
    (E) Amphibia

118. Which one of the following statements about cells is NOT true?

    (A) All living things are made up of one or more cells.
    (B) Cells may be seen with a microscope.
    (C) Cells are the basic units of life.
    (D) All cells come from pre-existing cells.
    (E) All cells have cell walls.

119. When refraction occurs, part of a wave

    (A) is bent more than another part
    (B) slows down before another part
    (C) is pushed to one side
    (D) is closer together than it appears to be
    (E) reflects back to the source

120. A bottle with a message is tossed into the ocean off the shore of Siberia in the Sea of Okhotsk. Where can it be expected to wash ashore?

    (A) Australia
    (B) Western North America
    (C) The Sea of Japan
    (D) Eastern Mexico
    (E) Western South America

PRACTICE TEST 2

Answer Key

1.	(A)	25.	(C)	49.	(D)	73.	(D)	97.	(C)
2.	(B)	26.	(A)	50.	(C)	74.	(E)	98.	(B)
3.	(C)	27.	(B)	51.	(C)	75.	(B)	99.	(A)
4.	(E)	28.	(B)	52.	(B)	76.	(A)	100.	(D)
5.	(C)	29.	(C)	53.	(B)	77.	(D)	101.	(A)
6.	(C)	30.	(B)	54.	(A)	78.	(C)	102.	(C)
7.	(E)	31.	(B)	55.	(A)	79.	(A)	103.	(C)
8.	(A)	32.	(B)	56.	(C)	80.	(C)	104.	(B)
9.	(B)	33.	(B)	57.	(A)	81.	(A)	105.	(B)
10.	(D)	34.	(B)	58.	(C)	82.	(D)	106.	(A)
11.	(B)	35.	(C)	59.	(D)	83.	(D)	107.	(B)
12.	(A)	36.	(E)	60.	(B)	84.	(D)	108.	(C)
13.	(C)	37.	(A)	61.	(B)	85.	(E)	109.	(D)
14.	(A)	38.	(B)	62.	(A)	86.	(D)	110.	(A)
15.	(C)	39.	(B)	63.	(A)	87.	(B)	111.	(E)
16.	(B)	40.	(C)	64.	(E)	88.	(B)	112.	(A)
17.	(E)	41.	(B)	65.	(A)	89.	(A)	113.	(B)
18.	(D)	42.	(D)	66.	(B)	90.	(D)	114.	(B)
19.	(D)	43.	(A)	67.	(A)	91.	(B)	115.	(D)
20.	(A)	44.	(B)	68.	(D)	92.	(B)	116.	(C)
21.	(B)	45.	(B)	69.	(B)	93.	(C)	117.	(C)
22.	(B)	46.	(A)	70.	(C)	94.	(C)	118.	(E)
23.	(B)	47.	(B)	71.	(C)	95.	(D)	119.	(B)
24.	(E)	48.	(B)	72.	(D)	96.	(C)	120.	(B)

PRACTICE TEST 2

Detailed Explanations of Answers

1. **(A)** Primary oocytes are typically present in great numbers in the female's ovaries at birth.

2. **(B)** All multicellular organisms that reproduce sexually begin life as a zygote. The zygote then undergoes a series of cell divisions known as cleavage. Cleavage is the division of the zygote cell.

3. **(C)** The primary spermatocytes then undergo meiosis I forming secondary spermatocytes with a single chromosome set. The secondary spermatocytes go through meiosis II, forming spermatid that are haploid. These spermatid then develop into the sperm cells.

4. **(E)** Primary oocytes undergo meiosis I forming one secondary oocyte and one smaller polar body. Both the secondary oocyte and the polar body undergo meiosis II—the polar body producing two polar bodies that are not functional cells, the oocyte producing one more polar body and one haploid egg cell.

5. **(C)** The secondary spermatocytes go through meiosis II, forming spermatid that are haploid.

6. **(C)** Positron decay results in the departure of a subatomic particle that carries away a positive charge and results in a proton being turned into a neutron. As a result, the mass number remains the same, but the atomic number decreases by 1.

7. **(E)** Gamma radiation is composed of photons of very short wavelength and high energy. Gamma radiation usually accompanies other types of radioactive decay and is the vehicle by which energy is carried away from the atom.

8. **(A)** Alpha decay results in a bundle of nuclear material leaving the atom. This is a very stable bundle of material that consists of two protons and two neutrons—equivalent to the nucleus of a helium atom.

DETAILED EXPLANATIONS OF ANSWERS | 241

9. **(B)** Beta decay is the opposite of positron decay and results in an electron carrying a negative charge away from the atom and ultimately turning a neutron into a proton. The overall mass number stays the same, but the atomic number increases by 1.

10. **(D)** Two electrons are transferred per atom in the oxidation-reduction reaction between magnesium and oxygen. Magnesium is oxidized by losing two electrons, and oxygen is reduced.

11. **(B)** The titration of acetic acid by a strong base results in the production of water and the acetate ion, which is a weak base—thereby leaving the solution basic.

12. **(A)** A cloudy product results in the formation of silver iodide (AgI) precipitate.

13. **(C)** In an electrolytic cell, melted sodium chloride will react by having the sodium ions move toward the cathode and the chloride ion move toward the anode. Sodium metal will be formed at the cathode, and chloride gas will be formed at the anode.

14. **(A)** The balance of the ecosystem can be destroyed by the removal, or decrease, of a single factor or by the addition, or increase, of a factor.

15. **(C)** Homeostasis is a dynamic balance that is achieved within an ecosystem that is functioning at its optimum level.

16. **(B)** The law of minimums (also known as the Liebig-Blackman Law of Limits) states that the resource in shortest supply in an ecosystem will limit population growth.

17. **(E)** The law of tolerance (or Shelford's law) states that population growth may be limited by having more of an element (such as heat or water) than it can tolerate.

18. **(D)** The number of individuals of a particular species that live in a particular area is known as the population density (number of organisms per area).

19. **(D)** Choices (A), (B), (C), and (E) are all evidence of homologous structures showing similar ancestry. However, choice (D), the bat, is an example of an analogous structure—similar but evolved from a different lineage for a similar purpose.

20. **(A)** F = ma. Multiply the mass (9.1 kg) by the acceleration (3.2 m/s²) to get 29.1 N.

21. **(B)** The lithosphere is composed primarily of molten magma. The lithosphere exists just below the crust.

22. **(B)** Logistic population growth is shown by a line that starts off flat, goes up quickly, and then flattens out again.

23. **(B)** Each level in an energy pyramid uses and/or loses as heat approximately 90% of the energy it begins with. This means that 10% of the energy of each level is available for use in the next level.

24. **(E)** Chordates include all organisms with spinal cords and vertebrae. Choices (A) through (D) are all invertebrates. A perch is a vertebrate and thus within the phylum Chordata.

25. **(C)** The spinal cord is not directly involved with immunity. The organs of the immune system in humans and other higher invertebrates include the lymph nodes, spleen, thymus, and tonsils.

26. **(A)** Oparin's hypothesis included the idea that most water on the Earth was in the form of steam, not ice. Oparin proposed that the Earth was approximately 4.6 billion years old and that there was a reducing atmosphere, meaning there was very little oxygen present. Instead there was an abundance of ammonia, hydrogen, methane, and steam (H_2O). The Earth was in the process of cooling down, so there was a great deal of heat energy available, as well as a pattern of recurring violent lightning storms providing another source of energy. During this cooling of the Earth, much of the steam surrounding the Earth would condense, forming hot seas. In the presence of abundant energy, the synthesis of simple organic molecules from the available chemicals became possible. These organic substances then collected in the hot, turbulent seas (sometimes referred to as the "primordial soup"). As the concentration of organic molecules became very high, they began forming into larger, charged complex molecules. Oparin called these highly absorptive molecules coacervates. Coacervates were also able to divide.

27. **(B)** The Gulf Stream consists of warm equatorial currents that raise the temperature of the winds that pass over England, thus warming the land. Being an island does not make England warmer. Latitude does affect temperature; the higher the latitude, the greater the Sun's angle and the cooler

the air. The winds that affect England do not come from France. Gulf Stream winds do not affect the Canadian coast.

28. **(B)** Answer choice (A) is one quark type too few; there are six of them. *Ceiling* and *floor* are not quark types. The six quark types are *up, down, strange, charm, bottom,* and *top* (or *beauty* and *truth* to European physicists). The three-quark combination for a proton is two *up* quarks and one *down* quark. The proton doesn't include a *strange* quark.

29. **(C)** The delocalized electrons in copper and other transition metals create an "electron sea," which is fundamental to the construction of the metallic attraction that holds solid copper together.

30. **(B)** The order of classification from least specific to most specific is kingdom, phylum, class, order, family, genus, and species. Of those listed, genus is the most specific, with members most closely related to each other.

31. **(B)** When one community completely replaces another over time in a given area, it is known as succession.

32. **(B)** The two oxygen atoms would produce two atoms of beryllium, each with four protons and four neutrons. The reaction doesn't form the same element since each of the resulting atoms contains only four protons, which identifies the element. Since the atom was split, it is no longer oxygen because each separate atom has four protons, not eight. Helium contains two protons, and neon contains ten.

33. **(B)** The strongest of the fundamental forces is the strong nuclear force that impacts objects at a range of about 10^{-15} meters. Then comes the electromagnetic force, which influences objects at an infinite range throughout the universe. After that, the weak nuclear force impacts objects that are in the same range as those impacted by the strong force. Gravity is the weakest of the four, but affects everything in the universe as well.

34. **(B)** Photosynthetic organisms are found in the Kingdom Plantae and the Kingdom Protista. Chordata is a phyla, not a kingdom.

35. **(C)** Although ΔT is a description of the change in temperature, it does not explain anything about why the water takes longer to boil than the piece of chocolate will to melt. A Btu is a unit of heat transfer that is quoted a lot in heating and cooling systems. It certainly is used to measure specific heat in units of Btu's per pound per degree Fahrenheit (Btu/lb/°F), but this

is a unit of measure, not a physical property of matter. Specific heat is the measurable ability of a substance to retain or lose heat. It is measured as the amount of heat required to raise the temperature of a gram of a substance by 1°C. Kelvin temperature is simply a measuring convention that relates all bodies to an absolute scale of heat transfer measurement, where zero is the temperature at which molecular motion ceases.

36. **(E)** The order of the layers of Earth's atmosphere from closest to farthest away is troposphere, stratosphere, mesosphere, thermosphere, and exosphere.

37. **(A)** Thermodynamics deals only in the large-scale responses of a system. It does not involve small-scale measurements, either alone or in combination with large-scale responses.

38. **(B)** Mitochondria are called the cell's "powerhouses," being the center of cellular respiration. (Cellular respiration is the process of breaking up covalent bonds within sugar molecules with the intake of oxygen and release of energy in the form of ATP [adenosine tri-phosphate] molecules. ATP is the energy form used by all cell processes.) Mitochondria (plural of mitochondrion) are found wherever energy is needed within the cell, and are more numerous in cells that require more energy (muscle, etc.).

39. **(B)** The size of a cell is limited by the ratio of its volume to its surface area. A cell will only remain stable if the surface area of the plasma membrane maintains a balance with the volume of cytoplasm.

40. **(C)** Earth acts through its center like a large magnet. Scientists are not sure what produces the enormous magnetic currents that are deep within the Earth and responsible for Earth's magnetic field. There is not enough magnetic material within the crust or mantle to explain the magnetic force of Earth based solely on the natural magnets. Turbulent motion in the upper core's electrically conductive fluid is suspected to be responsible for the magnetism, but a reasonable explanation is still lacking because of the complexity of how it operates, according to scientists who study it.

41. **(B)** When the pitch of a sound wave gets higher, that means that the wave is oscillating faster. This will happen when the source is moving toward the listener. If the source were moving away, the pitch would get lower. If the source were standing still, there would be no change in pitch. Moving the source up and down will not change its pitch. The amplitude of a wave affects the volume, but not the pitch.

42. **(D)** When the water pressure inside and outside the cell is equal, it is said to be in an isotonic (or isomotic) state. Water will pass through the membrane by osmosis from an area of higher concentration to an area of lower concentration (thus higher to lower water pressure) in order to produce isotonic conditions.

43. **(A)** According to Newton's first law of motion, the sleeping astronauts would continue their motion in a straight line if the shuttle suddenly changed velocity due to the firing of its thrusters. If the shuttle slowed down, the astronauts would move forward relative to the walls. This is all due to the fact that the astronauts have mass, or inertia. Gravity is so weak that it is negligible on the astronauts. They can be monitored by battery-driven devices that allow them to move around. They can certainly sleep without being strapped in. Bodies do not need a reaction force in order to sleep.

44. **(B)** While microtubules, microfilaments, and centrioles all provide structure to cells, cell walls are only found in plants. Mitochondria are found in both plant and animal cells, and are not used for structure, but for the processing of energy.

45. **(B)** Enzymes catalyze reactions in both living and nonliving environments.

46. **(A)** The proton's nemesis particle is the antiproton, which has the same mass but carries a negative charge. There is no such thing as an *antielectron;* the electron's antimatter is the *positron*. A *neutrino* is an extremely small elementary particle with no charge, that is created by certain types of radioactive decay. There is no such thing as a "negitron." However, Carl D. Anderson, who discovered the *positron,* did once consider changing the name of the electron to the "*negatron*" so that the term *electron* could be used as a term for a generically charged particle, either positive or negative. That did not happen, of course. That answer is also incorrect if the sequence of matter and antimatter is considered since the *positron* is the antimatter, and the matter would be the (correctly spelled) *negatron*. The neutrino and quark are both subatomic particles, not matter and antimatter.

47. **(B)** The AU is defined as the average distance of Earth from the Sun, which is approximately 150 million kilometers, or 93 million miles.

48. **(B)** Vitamins are organic cofactors or coenzymes that are required by some enzymatic reactions. In this case, it is the coenzyme required for the process of synthesizing collagen.

49. **(D)** Prosthetic groups, which may be ions or non-protein molecules, are similar to cofactors in that they facilitate the enzyme reaction. However, prosthetic groups are tightly attached by covalent bonds to the enzyme, rather than being separate atoms or molecules.

50. **(C)** An alluvial fan results when fast-moving mountain canyons transport large amounts of material out of the mountain into a slower-moving stream channel, where it is deposited into the bottom in a fan-shaped area. It differs from a delta in that its source is a canyon, whereas deltas are deposited at the base of a river into a lake or ocean. The tributary is the name of a smaller channel of water that feeds into a larger body of water. Topset beds are a type of delta sediment with coarse materials. A meandering stream is a single-channel stream that flows with a snake-like or serpentine pattern.

51. **(C)** *Adiabatic* comes from the Latin word meaning "does not pass through." In an adiabatic process involving a fluid, heat does not pass through the fluid. When heat passes between objects in a thermodynamic relationship, we call it a diabatic process, because there is a heat transfer. A common and practical example of an adiabatic situation is the effects of a thermal window pane. The air between the two panes of glass disallows the transfer of heat through the inside glass to the outer glass on a cold winter day.

52. **(B)** The egg-laying trait of monotremes is most likely an example of convergent evolution of that trait that is analogous to reptiles and birds. Analogous traits are an example of convergent evolution in which similar traits evolve in species without related ancestries.

53. **(B)** The ideal gas law assumes that there is no intermolecular attraction and that the volume of the gas molecules is insignificant relative to the space between the molecules. Therefore, gases that are large molecules and/or demonstrate strong intermolecular attraction are likely to be non-ideal gases. Hydrogen sulfide is both the largest gas molecule of the options and shows very strong intermolecular attraction, because of the difference in electronegativity between hydrogen and sulfur. Water vapor also shows strong intermolecular attraction, but it is a much smaller molecule.

54. **(A)** Einstein's equivalence theory explains how a person acting as an observer in a spacecraft enclosed in a completely sealed opaque chamber perceives the forces acting upon that spacecraft as if the person were experiencing the gravitational pull of Earth. This experience acts according to the equation, Force = (mass)(acceleration) or $F = ma$. The Michelson-

Morley experiment shows that the speed of light is the same in all directions, and has nothing to do with special relativity described here. Einstein's general theory of relativity involving the speed of light is not addressed in this question. This problem also has nothing to do with the Doppler effect.

55. **(A)** The process whereby cells build molecules and store energy (in the form of chemical bonds) is called anabolism.

56. **(C)** Cellular metabolism is a general term that includes all types of energy transformation processes, including photosynthesis, respiration, growth, movement, etc. These energy transformations occur as chemicals that are broken apart (catabolism) or synthesized within the cell (anabolism).

57. **(A)** Although (C) is technically correct, since on Earth mass and weight are the same value, the principle as expressed by Archimedes was stated in terms of weight of the object in a fluid.

58. **(C)** If the collection vessel is exposed to the surface of water, as it usually is when collecting a gas in the laboratory, then water evaporates from the surface and combines with the collected gas. Therefore, to be accurate in calculating how many moles of gas are produced in the reaction, Dalton's law must be used to account for the moles of water vapor also in the collection vessel. Dalton's law states that the total pressure exerted by a mixture of gases equals the sum of the partial pressures of each component in the mixture.

59. **(D)** An electromagnet is not a permanent magnet; it ceases to be a magnet when the electricity ceases to flow through the wire. The strength of an electromagnet can be increased easily by increasing the number of coils in the wire. Since choice (D) allows the selection of choice (B), which states that an electromagnet is permanent, and choice (C), which means that the strength of a magnet is limited, choice (D) includes both incorrect answers, making it the correct answer. The first answer is a true statement; therefore, it is not the correct answer. Choice (E) includes choice (A), so it should not be selected.

60. **(B)** About 500 volcanoes are considered to be active on Earth, with two-thirds of them in the so-called ring of fire, which extends around the Pacific Ocean.

61. **(B)** P-waves, also called push waves, travel out directly from the epicenter of an earthquake and form longitudinal waves. The other answers represent transverse waves. Gravity is not a wave; it is a force.

62. **(A)** Based on specific gravity alone, an object that is denser than water will sink. A substance will not dissolve in water or increase water temperature based upon its specific gravity, though it may do so because of other properties. The ability of a substance to mix with water is not due to its specific gravity.

63. **(A)** The tide is the continuous cycle of an alternatively rising and falling sea level along the coastlines and in bodies of water connected to the sea. The Moon is the single greatest influence on the tides.

64. **(E)** The illustration is called a Punnett square.

65. **(A)** The genotypes of the parents are TT and tt.

66. **(B)** Every gamete will have either the T gene or the t gene.

67. **(A)** All offspring will have a genotype Tt and a dominant gene for tallness (T), so all will have a phenotype of tall. Thus, the ratio is 4 tall; 0 short.

68. **(D)** The genotype ratio will be 1 TT: 2 Tt: 1 tt. Since T (tall) is dominant to t (short), only one out of four will be short. The phenotypic ratio will be 3 tall; 1 short.

69. **(B)** The mass of the object does not change because it is measured independently of the force of gravity on Earth or acceleration of the spacecraft. This is what differentiates mass from weight.

70. **(C)** The angle through which the star appears to move in the course of half a year is called the parallax. The time involved is based on measuring the angle of observation from both extremes of Earth's orbit around the Sun, which is at a maximum separation every half year.

71. **(C)** Kingdom Monera contains prokaryotic cells, which are considered to be the first to evolve.

72. **(D)** Lengths have to do with intervals of space, measured in three dimensions. The standard of measurement in physics is the meter. Mass is the amount of substance that an object contains and is measured in grams or kilograms. Time is a measurement of duration that we measure in seconds. The system of units most commonly used by scientists is the m-k-s system; astrophysics measures in centimeters, grams and seconds, known as the c-g-s system.

73. **(D)** The highest melting point will be demonstrated by the crystal that has the strongest forces holding it in the crystalline structure. In this case,

SiC is held together with network covalent attractions, which are stronger than ionic and polar intermolecular attractions.

74. **(E)** In the center of the root is the vascular cylinder (e), including xylem and phloem tissue.

75. **(B)** The primary tissues include the outermost layer, the epidermis (b). The epidermis is one cell layer thick, and serves to protect the internal root tissue and absorb nutrients and water.

76. **(A)** In the maturation region, the epidermis produces root hairs (a), extensions of the cells that reach between soil particles and retrieve water and minerals.

77. **(D)** Inside the cortex is a ring of endodermis (d), a single layer of cells that are tightly connected, so no substances can pass between cells. This feature allows the endodermis to act as a filter; all substances entering the vascular tissues from the root must pass through these cells.

78. **(C)** Inside the epidermis is a ring known as the cortex (c), made up of large parenchyma cells. Parenchyma cells are present in many tissues of plants; they are thin-walled cells loosely packed to allow for the flow of gases and mineral uptake.

79. **(A)** The photosphere is what we observe on the Sun's surface. Plasma is the state of matter engulfing the Sun's interior, and represents the fourth state of matter. A sunspot and a flare are local phenomena occurring on the Sun's surface. The corona is the outside ring of the Sun's surface which is visible during an eclipse.

80. **(C)** Considering the size of an object does not describe it quantitatively. Mass is correct but density represents mass per unit volume, which is not needed to find weight. The force it exerts is not upward but downward toward the Earth by gravity. Weight and mass are both able to be determined independently of the type of material it contains. The amount of pressure exerted on an object does not determine its weight, but rather the amount of gravity exerted does.

81. **(A)** The cell membrane is composed of a double layer (bilayer) of phospholipids with protein globules imbedded within the layers. The construction of the membrane allows it to aid cell function by permitting entrance and exit of molecules as needed by the cell.

82. **(D)** If the distance from the source of energy doubles, the brightness reduces by a factor of x^2. Square the distance four times to get 4^2 or 16. So the star is 16 times dimmer at 4 times the distance away.

83. **(D)** Oxygen is carried by hemoglobin molecules (that contain iron) in red blood cells. Lymphocytes and erythrocytes are types of blood cells; carbohydrates are sugar molecules; and gamma globulins are protein molecules found in the blood.

84. **(D)** Dicots (plants with two cotyledons in each seed) have taproot systems and flowers with petals in multiples of four or five. The leaves of a dicot will have networked or branching veins, and the vascular bundles within the stem will be arranged in rings.

85. **(E)** Light behaves as both a particle and a wave. The speed of light is a constant value of 3.0×10^8 m/s or 183,310 mi/s. Prior to Einstein's theories, answer (D) was the common scientific theory that Einstein helped to disprove.

86. **(D)** Lightning heats the air to several thousand degrees in only a fraction of a second. This causes the air to expand rapidly. Surrounding air rushes in to fill the space left by the expanding heated air. It is the air rushing in which creates the sonic boom we know as thunder. By now, it should be evident that lightning and thunder originate at essentially the same location. In addition, the energy released by the lightning creates the noise associated with the event. Some thunder rumbles for a very long time. There may actually be no rain at all when the lightning strikes; the sound would speed up if it travels through a liquid—not slow down. Light travels at 3×10^8 m/s; sound travels at a speed of 343 m/s in air, which completely explains why we see the lightning before hearing the thunder.

87. **(B)** Temporary movement into a new range and back again is known as migration.

88. **(B)** To calculate the specific heat, we need to determine the amount of calories transferred per degree Celsius. However, since we have 6 grams of the substance, we must divide the 12 calories by 6 grams, which comes out to 2 calories per gram (rounded to three significant digits). We then divide by the temperature of 2.345°C to get the specific heat of .852 cal/g/°C. We would erroneously get (A) if we neglect to divide by the increase in temperature. Answer (C) is reached erroneously by dividing the temperature by

the calories per gram. Choices (D) and (E) are inaccurately reached if we neglect to divide the calories.

89. **(A)** Einstein's equivalence theory explains how a person acting as an observer in a spacecraft enclosed in a chamber that is completely sealed and opaque perceives the forces acting upon that spacecraft in the same manner as if the person were experiencing the gravitational pull of Earth according to the equation, Force = (mass)(acceleration) or $F = ma$. In this case, the absence of gravity is due to the distance above Earth that the space shuttle is orbiting. The Michelson-Morley experiment shows that the speed of light is the same in all directions and has nothing to do with special relativity as described here. Einstein's general theory of relativity involving the speed of light is expressed in the equation $E = mc^2$, which again is not addressed in this question. Although answer choice (E) mentions special relativity, it does so in terms of velocity rather than acceleration, which is erroneous.

90. **(D)** The intensity of the lightning is determined by complex processes which electrify the cloud. The updraft is only indirectly associated with this. The height of the cloud base is determined by the vertical temperature and dew point lapse rates in the atmosphere. This height can be estimated well before a thunderstorm ever forms and is not dependent on the updraft. Wind gusts at the surface are mainly caused by the downdraft. The speed of rain droplets is dependent on gravity and is unaffected by updraft. It is the strength of the updraft that determines how large individual precipitation elements will grow before they can no longer remain suspended in the cloud. Thus, stronger updrafts produce larger raindrops (or hailstones) than do weaker ones.

91. **(B)** The reproductive organs of the sporophyte generation produce gametophytes through the process of meiosis. Gametophytes may be male or female and are haploid. Either can be considered first or second generations. Zygotes are produced when the spore fertilizes the egg.

92. **(B)** Since an atom of carbon must have six protons, by definition, in order to be carbon, then the atomic mass of the isotope is the number of protons and neutrons in its nucleus (6 + 6 = 12).

93. **(C)** Magnesium demonstrates the greatest difference between its second and third ionization energies because it has two electrons in its outer shell. To remove the third would require breaking into the inner, more tightly held second shell—which requires considerably more energy.

94. **(C)** The male structure is the stamen, consisting of the anther atop a long, hollow filament. The anther has four lobes and contains the cells (microspore mother cells) that become pollen. Some mature pollen grains are conveyed (usually by wind, birds, or insects) to a flower of a compatible species where they stick to the stigma. The stigma produces chemicals that stimulate the pollen to burrow into the style, forming a hollow pollen tube. This tube is produced by the tube nucleus that has developed from a portion of the pollen grain. The pollen tube extends down toward the ovary.

95. **(D)** The universe is infinite; there is nothing outside of it. When infinity expands, it is still infinity, just bigger. Our telescopes are able to see to the ends of the perceived universe. We don't see more matter at the ends of the universe. We see space expanding and matter ending. There is no evidence that the universe is spherical or that it will be sucked into a huge black hole. But it sure makes for a great sci-fi movie!

96. **(C)** The core is speculated to contain a liquid center within the inner core. The other layers of Earth are all solid.

97. **(C)** The first law of Mendelian genetics is the law of segregation, stating that traits are expressed from a pair of genes in the individual, one of which came from each parent. The genes are separated in gamete formation, and then randomly brought together in fertilization.

98. **(B)** To convert from Celsius to Fahrenheit, use the formula $C = 5/9(F - 32)$. Plugging in the numbers will yield $C = 5/9 (54.5 - 32) = 5/9(22.5) = 12.5°C$. Errors common in the other answers are because of either using the wrong formula, or incorrectly manipulating the numbers mathematically.

99. **(A)** The mass of the cathode will increase because the copper ions in solution will gain electrons and precipitate solid copper onto the cathode.

100. **(D)** Epithelial (skin) cells have no direct function in immunity. Cells involved in immunity are called lymphocytes and begin in bone marrow as stem cells. There are two classes of lymphocytes, B cells and T cells. B cells emerge from the bone marrow, mature, and produce antibodies into the bloodstream that find and attach themselves to foreign antigens (toxins, bacteria, foreign cells, etc.). The attachment of an antibody to an antigen marks the pair for destruction. T cells mature in the thymus gland. Some T cells patrol the blood for antigens like B cells, but T cells are also equipped to destroy antigens themselves. T cells also regulate the body's immune responses.

101. **(A)** All of the items except pond water temperature will directly affect the snake's niche. Hawks are predators of snakes, preventing the snake population from being overgrown. Mice are the main food source, so without them the snakes will suffer. Daily temperature and yearly rainfall amounts determine the types of plants that will grow and provide shade and camouflage, as well as the body temperature of the snake (needs warm climate). An organism's niche is the role it plays within the ecosystem. It includes its physical requirements (such as light and water) and its biological activities (how it reproduces, how it acquires food etc.). In most cases, the most important aspect of an organism's niche is its place in the food chain.

102. **(C)** To be considered a mineral, the substance must never have been part of a living organism and must be found in nature. Minerals are the most common form of solid material found in Earth's crust.

103. **(C)** All of the choices are true except (C). Animal species are capable of sexual reproduction, though some are also capable of asexual reproduction (e.g., hydra).

104. **(B)** Neither plants nor animals are able to use nitrogen directly from the air. Instead, a process known as nitrogen fixation makes nitrogen available for absorption by the roots of plants. Nitrogen fixation is the process of combining it with either hydrogen or oxygen. Nitrogen fixation is accomplished in one of two ways—either by nitrogen-fixing bacteria or by the action of lightning.

105. **(B)** The conductivity of the cement allows heat from your fingers to flow to it more than the wood does. This heat flow feels as if the cement is cooler than the wood even though both floors may be the same temperature.

106. **(A)** Moving to the right on the periodic table across a row, each new element experiences a greater pull between the nucleus and the ever-increasing number of electrons, which pulls in the outer shell of electrons and decreases the atomic radius.

107. **(B)** A decrease of one unit in magnitude corresponds to an increase of brightness by a factor of 2.5. Since Star X appears to be 2.5 times as bright as Star Y, their magnitudes differ by a value of 1 unit.

108. **(C)** Animal respiration releases carbon dioxide back into the atmosphere in large quantities; it does not take in carbon dioxide for use. Most of the carbon within organisms is derived from the production of carbohydrates

in plants through photosynthesis. Carbon is also dissolved directly into the oceans, where it is combined with calcium to form calcium carbonate—used by mollusks to form their shells. Detritus feeders include worms, mites, insects, and crustaceans that feed on dead organic matter, returning the carbon to the cycle through chemical breakdown and respiration. Organic matter that is left to decay may, under conditions of heat and pressure, be transformed into coal, oil, or natural gas—the fossil fuels. When fossil fuels are burned for energy, the combustion process releases carbon dioxide back into the atmosphere, where it is available to plants for photosynthesis.

109. **(D)** The stomach does not secrete acetic acid.

110. **(A)** Firmness is not a measureable physical property of minerals, so it is the correct answer to this question. Hardness is another one not listed in this question. The others are all identifiable properties of minerals.

111. **(E)** The selective permeability of the cell membrane serves to manage the concentration of substances within the cell, preserving its health. There are two methods by which substances can cross the cell membrane: passive transport and active transport.

112. **(A)** The ecliptic is the apparent annual path "through the planets and stars" that the Sun follows. The celestial equator is the imaginary line surrounding Earth that extends the equator up into the heavens. The equinox is not an astronomical boundary, but represents the time of year when the amount of daylight and nightlight are equal in spring and fall. The solstice is the same calendar demarcation in winter and summer, when the shortest and longest daylight occurs at the start of these seasons, and Earth is closest and farthest from the Sun during its annual trajectory. An orbital path is simply that—the generic name for the path that a satellite, either natural or artificial, takes as it travels around another celestial body.

113. **(B)** A species' (and thus an organism's) habitat must include all the factors that will support its life and reproduction. These factors may be biotic (i.e., living—food source, predators, etc.) or abiotic (i.e., nonliving—weather, temperature, soil features, etc.).

114. **(B)** CH_4 demonstrates 109.5° between the C-H bonds, while XeF_4 demonstrates 90° angles between its Xe-F bonds. Even though the formulas are similar, the different geometry is a result of the larger size of the Xe atom and the highly electronegative F atoms creating an "expanded octet." Two

other pairs of unbonded electron pairs around Xenon also take up space, forcing the bonds to fluorine to be closer together than the C-H bonds in methane.

115. **(D)** Heat of vaporization (H), or enthalpy, is the amount of energy needed to turn a given quantity of a substance into a gas. Heat of fusion is the change of a solid to a liquid or vice versa. A heat index has nothing to do with changes of state. It is a measure of the amount of heat in a particular environmental area. A heat wave constant is a fictitious term. Specific heat is the amount of heat energy necessary to raise 1g of a substance by 1°C.

116. **(C)** The sharp boundary of a community is called an ecotone.

117. **(C)** Bony fish belong to the class Osteichthyes. Since soft tissues would not fossilize well, it would be unlikely to find a fossil of a cartilaginous (non-bony) fish (Chondrichthyes). Reptilia, Aves, and Amphibia are not classes of fish.

118. **(E)** Only plant cells have cell walls, while all cells have a cell/plasma membrane. The rest of the choices are tenets of cell theory.

119. **(B)** Refraction is the bending of waves toward the direction of the slower wave velocity. One part of the wave slows down before another part. One part of the wave does not bend more than another part; it is not pushed to one side but goes through the medium equally. It is also not closer together at one part than another. The last answer explains reflection, not refraction.

120. **(B)** The bottle will drift eastward until it will ride the Kuroshio current to the North Pacific Drift and wash ashore on the west coast of North America.

PRACTICE TEST 1

Answer Sheet

1. Ⓐ Ⓑ Ⓒ Ⓓ Ⓔ
2. Ⓐ Ⓑ Ⓒ Ⓓ Ⓔ
3. Ⓐ Ⓑ Ⓒ Ⓓ Ⓔ
4. Ⓐ Ⓑ Ⓒ Ⓓ Ⓔ
5. Ⓐ Ⓑ Ⓒ Ⓓ Ⓔ
6. Ⓐ Ⓑ Ⓒ Ⓓ Ⓔ
7. Ⓐ Ⓑ Ⓒ Ⓓ Ⓔ
8. Ⓐ Ⓑ Ⓒ Ⓓ Ⓔ
9. Ⓐ Ⓑ Ⓒ Ⓓ Ⓔ
10. Ⓐ Ⓑ Ⓒ Ⓓ Ⓔ
11. Ⓐ Ⓑ Ⓒ Ⓓ Ⓔ
12. Ⓐ Ⓑ Ⓒ Ⓓ Ⓔ
13. Ⓐ Ⓑ Ⓒ Ⓓ Ⓔ
14. Ⓐ Ⓑ Ⓒ Ⓓ Ⓔ
15. Ⓐ Ⓑ Ⓒ Ⓓ Ⓔ
16. Ⓐ Ⓑ Ⓒ Ⓓ Ⓔ
17. Ⓐ Ⓑ Ⓒ Ⓓ Ⓔ
18. Ⓐ Ⓑ Ⓒ Ⓓ Ⓔ
19. Ⓐ Ⓑ Ⓒ Ⓓ Ⓔ
20. Ⓐ Ⓑ Ⓒ Ⓓ Ⓔ
21. Ⓐ Ⓑ Ⓒ Ⓓ Ⓔ
22. Ⓐ Ⓑ Ⓒ Ⓓ Ⓔ
23. Ⓐ Ⓑ Ⓒ Ⓓ Ⓔ
24. Ⓐ Ⓑ Ⓒ Ⓓ Ⓔ
25. Ⓐ Ⓑ Ⓒ Ⓓ Ⓔ
26. Ⓐ Ⓑ Ⓒ Ⓓ Ⓔ
27. Ⓐ Ⓑ Ⓒ Ⓓ Ⓔ
28. Ⓐ Ⓑ Ⓒ Ⓓ Ⓔ
29. Ⓐ Ⓑ Ⓒ Ⓓ Ⓔ
30. Ⓐ Ⓑ Ⓒ Ⓓ Ⓔ
31. Ⓐ Ⓑ Ⓒ Ⓓ Ⓔ
32. Ⓐ Ⓑ Ⓒ Ⓓ Ⓔ
33. Ⓐ Ⓑ Ⓒ Ⓓ Ⓔ
34. Ⓐ Ⓑ Ⓒ Ⓓ Ⓔ
35. Ⓐ Ⓑ Ⓒ Ⓓ Ⓔ
36. Ⓐ Ⓑ Ⓒ Ⓓ Ⓔ
37. Ⓐ Ⓑ Ⓒ Ⓓ Ⓔ
38. Ⓐ Ⓑ Ⓒ Ⓓ Ⓔ
39. Ⓐ Ⓑ Ⓒ Ⓓ Ⓔ
40. Ⓐ Ⓑ Ⓒ Ⓓ Ⓔ
41. Ⓐ Ⓑ Ⓒ Ⓓ Ⓔ
42. Ⓐ Ⓑ Ⓒ Ⓓ Ⓔ
43. Ⓐ Ⓑ Ⓒ Ⓓ Ⓔ
44. Ⓐ Ⓑ Ⓒ Ⓓ Ⓔ
45. Ⓐ Ⓑ Ⓒ Ⓓ Ⓔ
46. Ⓐ Ⓑ Ⓒ Ⓓ Ⓔ
47. Ⓐ Ⓑ Ⓒ Ⓓ Ⓔ
48. Ⓐ Ⓑ Ⓒ Ⓓ Ⓔ
49. Ⓐ Ⓑ Ⓒ Ⓓ Ⓔ
50. Ⓐ Ⓑ Ⓒ Ⓓ Ⓔ
51. Ⓐ Ⓑ Ⓒ Ⓓ Ⓔ
52. Ⓐ Ⓑ Ⓒ Ⓓ Ⓔ
53. Ⓐ Ⓑ Ⓒ Ⓓ Ⓔ
54. Ⓐ Ⓑ Ⓒ Ⓓ Ⓔ
55. Ⓐ Ⓑ Ⓒ Ⓓ Ⓔ
56. Ⓐ Ⓑ Ⓒ Ⓓ Ⓔ
57. Ⓐ Ⓑ Ⓒ Ⓓ Ⓔ
58. Ⓐ Ⓑ Ⓒ Ⓓ Ⓔ
59. Ⓐ Ⓑ Ⓒ Ⓓ Ⓔ
60. Ⓐ Ⓑ Ⓒ Ⓓ Ⓔ
61. Ⓐ Ⓑ Ⓒ Ⓓ Ⓔ
62. Ⓐ Ⓑ Ⓒ Ⓓ Ⓔ
63. Ⓐ Ⓑ Ⓒ Ⓓ Ⓔ
64. Ⓐ Ⓑ Ⓒ Ⓓ Ⓔ
65. Ⓐ Ⓑ Ⓒ Ⓓ Ⓔ
66. Ⓐ Ⓑ Ⓒ Ⓓ Ⓔ
67. Ⓐ Ⓑ Ⓒ Ⓓ Ⓔ
68. Ⓐ Ⓑ Ⓒ Ⓓ Ⓔ
69. Ⓐ Ⓑ Ⓒ Ⓓ Ⓔ
70. Ⓐ Ⓑ Ⓒ Ⓓ Ⓔ
71. Ⓐ Ⓑ Ⓒ Ⓓ Ⓔ
72. Ⓐ Ⓑ Ⓒ Ⓓ Ⓔ
73. Ⓐ Ⓑ Ⓒ Ⓓ Ⓔ
74. Ⓐ Ⓑ Ⓒ Ⓓ Ⓔ
75. Ⓐ Ⓑ Ⓒ Ⓓ Ⓔ
76. Ⓐ Ⓑ Ⓒ Ⓓ Ⓔ
77. Ⓐ Ⓑ Ⓒ Ⓓ Ⓔ
78. Ⓐ Ⓑ Ⓒ Ⓓ Ⓔ

(Continued)

PRACTICE TEST 1

Answer Sheet

79. Ⓐ Ⓑ Ⓒ Ⓓ Ⓔ
80. Ⓐ Ⓑ Ⓒ Ⓓ Ⓔ
81. Ⓐ Ⓑ Ⓒ Ⓓ Ⓔ
82. Ⓐ Ⓑ Ⓒ Ⓓ Ⓔ
83. Ⓐ Ⓑ Ⓒ Ⓓ Ⓔ
84. Ⓐ Ⓑ Ⓒ Ⓓ Ⓔ
85. Ⓐ Ⓑ Ⓒ Ⓓ Ⓔ
86. Ⓐ Ⓑ Ⓒ Ⓓ Ⓔ
87. Ⓐ Ⓑ Ⓒ Ⓓ Ⓔ
88. Ⓐ Ⓑ Ⓒ Ⓓ Ⓔ
89. Ⓐ Ⓑ Ⓒ Ⓓ Ⓔ
90. Ⓐ Ⓑ Ⓒ Ⓓ Ⓔ
91. Ⓐ Ⓑ Ⓒ Ⓓ Ⓔ
92. Ⓐ Ⓑ Ⓒ Ⓓ Ⓔ

93. Ⓐ Ⓑ Ⓒ Ⓓ Ⓔ
94. Ⓐ Ⓑ Ⓒ Ⓓ Ⓔ
95. Ⓐ Ⓑ Ⓒ Ⓓ Ⓔ
96. Ⓐ Ⓑ Ⓒ Ⓓ Ⓔ
97. Ⓐ Ⓑ Ⓒ Ⓓ Ⓔ
98. Ⓐ Ⓑ Ⓒ Ⓓ Ⓔ
99. Ⓐ Ⓑ Ⓒ Ⓓ Ⓔ
100. Ⓐ Ⓑ Ⓒ Ⓓ Ⓔ
101. Ⓐ Ⓑ Ⓒ Ⓓ Ⓔ
102. Ⓐ Ⓑ Ⓒ Ⓓ Ⓔ
103. Ⓐ Ⓑ Ⓒ Ⓓ Ⓔ
104. Ⓐ Ⓑ Ⓒ Ⓓ Ⓔ
105. Ⓐ Ⓑ Ⓒ Ⓓ Ⓔ
106. Ⓐ Ⓑ Ⓒ Ⓓ Ⓔ

107. Ⓐ Ⓑ Ⓒ Ⓓ Ⓔ
108. Ⓐ Ⓑ Ⓒ Ⓓ Ⓔ
109. Ⓐ Ⓑ Ⓒ Ⓓ Ⓔ
110. Ⓐ Ⓑ Ⓒ Ⓓ Ⓔ
111. Ⓐ Ⓑ Ⓒ Ⓓ Ⓔ
112. Ⓐ Ⓑ Ⓒ Ⓓ Ⓔ
113. Ⓐ Ⓑ Ⓒ Ⓓ Ⓔ
114. Ⓐ Ⓑ Ⓒ Ⓓ Ⓔ
115. Ⓐ Ⓑ Ⓒ Ⓓ Ⓔ
116. Ⓐ Ⓑ Ⓒ Ⓓ Ⓔ
117. Ⓐ Ⓑ Ⓒ Ⓓ Ⓔ
118. Ⓐ Ⓑ Ⓒ Ⓓ Ⓔ
119. Ⓐ Ⓑ Ⓒ Ⓓ Ⓔ
120. Ⓐ Ⓑ Ⓒ Ⓓ Ⓔ

PRACTICE TEST 2

Answer Sheet

1. Ⓐ Ⓑ Ⓒ Ⓓ Ⓔ
2. Ⓐ Ⓑ Ⓒ Ⓓ Ⓔ
3. Ⓐ Ⓑ Ⓒ Ⓓ Ⓔ
4. Ⓐ Ⓑ Ⓒ Ⓓ Ⓔ
5. Ⓐ Ⓑ Ⓒ Ⓓ Ⓔ
6. Ⓐ Ⓑ Ⓒ Ⓓ Ⓔ
7. Ⓐ Ⓑ Ⓒ Ⓓ Ⓔ
8. Ⓐ Ⓑ Ⓒ Ⓓ Ⓔ
9. Ⓐ Ⓑ Ⓒ Ⓓ Ⓔ
10. Ⓐ Ⓑ Ⓒ Ⓓ Ⓔ
11. Ⓐ Ⓑ Ⓒ Ⓓ Ⓔ
12. Ⓐ Ⓑ Ⓒ Ⓓ Ⓔ
13. Ⓐ Ⓑ Ⓒ Ⓓ Ⓔ
14. Ⓐ Ⓑ Ⓒ Ⓓ Ⓔ
15. Ⓐ Ⓑ Ⓒ Ⓓ Ⓔ
16. Ⓐ Ⓑ Ⓒ Ⓓ Ⓔ
17. Ⓐ Ⓑ Ⓒ Ⓓ Ⓔ
18. Ⓐ Ⓑ Ⓒ Ⓓ Ⓔ
19. Ⓐ Ⓑ Ⓒ Ⓓ Ⓔ
20. Ⓐ Ⓑ Ⓒ Ⓓ Ⓔ
21. Ⓐ Ⓑ Ⓒ Ⓓ Ⓔ
22. Ⓐ Ⓑ Ⓒ Ⓓ Ⓔ
23. Ⓐ Ⓑ Ⓒ Ⓓ Ⓔ
24. Ⓐ Ⓑ Ⓒ Ⓓ Ⓔ
25. Ⓐ Ⓑ Ⓒ Ⓓ Ⓔ
26. Ⓐ Ⓑ Ⓒ Ⓓ Ⓔ
27. Ⓐ Ⓑ Ⓒ Ⓓ Ⓔ
28. Ⓐ Ⓑ Ⓒ Ⓓ Ⓔ
29. Ⓐ Ⓑ Ⓒ Ⓓ Ⓔ
30. Ⓐ Ⓑ Ⓒ Ⓓ Ⓔ
31. Ⓐ Ⓑ Ⓒ Ⓓ Ⓔ
32. Ⓐ Ⓑ Ⓒ Ⓓ Ⓔ
33. Ⓐ Ⓑ Ⓒ Ⓓ Ⓔ
34. Ⓐ Ⓑ Ⓒ Ⓓ Ⓔ
35. Ⓐ Ⓑ Ⓒ Ⓓ Ⓔ
36. Ⓐ Ⓑ Ⓒ Ⓓ Ⓔ
37. Ⓐ Ⓑ Ⓒ Ⓓ Ⓔ
38. Ⓐ Ⓑ Ⓒ Ⓓ Ⓔ
39. Ⓐ Ⓑ Ⓒ Ⓓ Ⓔ
40. Ⓐ Ⓑ Ⓒ Ⓓ Ⓔ
41. Ⓐ Ⓑ Ⓒ Ⓓ Ⓔ
42. Ⓐ Ⓑ Ⓒ Ⓓ Ⓔ
43. Ⓐ Ⓑ Ⓒ Ⓓ Ⓔ
44. Ⓐ Ⓑ Ⓒ Ⓓ Ⓔ
45. Ⓐ Ⓑ Ⓒ Ⓓ Ⓔ
46. Ⓐ Ⓑ Ⓒ Ⓓ Ⓔ
47. Ⓐ Ⓑ Ⓒ Ⓓ Ⓔ
48. Ⓐ Ⓑ Ⓒ Ⓓ Ⓔ
49. Ⓐ Ⓑ Ⓒ Ⓓ Ⓔ
50. Ⓐ Ⓑ Ⓒ Ⓓ Ⓔ
51. Ⓐ Ⓑ Ⓒ Ⓓ Ⓔ
52. Ⓐ Ⓑ Ⓒ Ⓓ Ⓔ
53. Ⓐ Ⓑ Ⓒ Ⓓ Ⓔ
54. Ⓐ Ⓑ Ⓒ Ⓓ Ⓔ
55. Ⓐ Ⓑ Ⓒ Ⓓ Ⓔ
56. Ⓐ Ⓑ Ⓒ Ⓓ Ⓔ
57. Ⓐ Ⓑ Ⓒ Ⓓ Ⓔ
58. Ⓐ Ⓑ Ⓒ Ⓓ Ⓔ
59. Ⓐ Ⓑ Ⓒ Ⓓ Ⓔ
60. Ⓐ Ⓑ Ⓒ Ⓓ Ⓔ
61. Ⓐ Ⓑ Ⓒ Ⓓ Ⓔ
62. Ⓐ Ⓑ Ⓒ Ⓓ Ⓔ
63. Ⓐ Ⓑ Ⓒ Ⓓ Ⓔ
64. Ⓐ Ⓑ Ⓒ Ⓓ Ⓔ
65. Ⓐ Ⓑ Ⓒ Ⓓ Ⓔ
66. Ⓐ Ⓑ Ⓒ Ⓓ Ⓔ
67. Ⓐ Ⓑ Ⓒ Ⓓ Ⓔ
68. Ⓐ Ⓑ Ⓒ Ⓓ Ⓔ
69. Ⓐ Ⓑ Ⓒ Ⓓ Ⓔ
70. Ⓐ Ⓑ Ⓒ Ⓓ Ⓔ
71. Ⓐ Ⓑ Ⓒ Ⓓ Ⓔ
72. Ⓐ Ⓑ Ⓒ Ⓓ Ⓔ
73. Ⓐ Ⓑ Ⓒ Ⓓ Ⓔ
74. Ⓐ Ⓑ Ⓒ Ⓓ Ⓔ
75. Ⓐ Ⓑ Ⓒ Ⓓ Ⓔ
76. Ⓐ Ⓑ Ⓒ Ⓓ Ⓔ
77. Ⓐ Ⓑ Ⓒ Ⓓ Ⓔ
78. Ⓐ Ⓑ Ⓒ Ⓓ Ⓔ

(Continued)

PRACTICE TEST 2

Answer Sheet

79. Ⓐ Ⓑ Ⓒ Ⓓ Ⓔ
80. Ⓐ Ⓑ Ⓒ Ⓓ Ⓔ
81. Ⓐ Ⓑ Ⓒ Ⓓ Ⓔ
82. Ⓐ Ⓑ Ⓒ Ⓓ Ⓔ
83. Ⓐ Ⓑ Ⓒ Ⓓ Ⓔ
84. Ⓐ Ⓑ Ⓒ Ⓓ Ⓔ
85. Ⓐ Ⓑ Ⓒ Ⓓ Ⓔ
86. Ⓐ Ⓑ Ⓒ Ⓓ Ⓔ
87. Ⓐ Ⓑ Ⓒ Ⓓ Ⓔ
88. Ⓐ Ⓑ Ⓒ Ⓓ Ⓔ
89. Ⓐ Ⓑ Ⓒ Ⓓ Ⓔ
90. Ⓐ Ⓑ Ⓒ Ⓓ Ⓔ
91. Ⓐ Ⓑ Ⓒ Ⓓ Ⓔ
92. Ⓐ Ⓑ Ⓒ Ⓓ Ⓔ

93. Ⓐ Ⓑ Ⓒ Ⓓ Ⓔ
94. Ⓐ Ⓑ Ⓒ Ⓓ Ⓔ
95. Ⓐ Ⓑ Ⓒ Ⓓ Ⓔ
96. Ⓐ Ⓑ Ⓒ Ⓓ Ⓔ
97. Ⓐ Ⓑ Ⓒ Ⓓ Ⓔ
98. Ⓐ Ⓑ Ⓒ Ⓓ Ⓔ
99. Ⓐ Ⓑ Ⓒ Ⓓ Ⓔ
100. Ⓐ Ⓑ Ⓒ Ⓓ Ⓔ
101. Ⓐ Ⓑ Ⓒ Ⓓ Ⓔ
102. Ⓐ Ⓑ Ⓒ Ⓓ Ⓔ
103. Ⓐ Ⓑ Ⓒ Ⓓ Ⓔ
104. Ⓐ Ⓑ Ⓒ Ⓓ Ⓔ
105. Ⓐ Ⓑ Ⓒ Ⓓ Ⓔ
106. Ⓐ Ⓑ Ⓒ Ⓓ Ⓔ

107. Ⓐ Ⓑ Ⓒ Ⓓ Ⓔ
108. Ⓐ Ⓑ Ⓒ Ⓓ Ⓔ
109. Ⓐ Ⓑ Ⓒ Ⓓ Ⓔ
110. Ⓐ Ⓑ Ⓒ Ⓓ Ⓔ
111. Ⓐ Ⓑ Ⓒ Ⓓ Ⓔ
112. Ⓐ Ⓑ Ⓒ Ⓓ Ⓔ
113. Ⓐ Ⓑ Ⓒ Ⓓ Ⓔ
114. Ⓐ Ⓑ Ⓒ Ⓓ Ⓔ
115. Ⓐ Ⓑ Ⓒ Ⓓ Ⓔ
116. Ⓐ Ⓑ Ⓒ Ⓓ Ⓔ
117. Ⓐ Ⓑ Ⓒ Ⓓ Ⓔ
118. Ⓐ Ⓑ Ⓒ Ⓓ Ⓔ
119. Ⓐ Ⓑ Ⓒ Ⓓ Ⓔ
120. Ⓐ Ⓑ Ⓒ Ⓓ Ⓔ

Glossary

absolute zero: theoretical temperature at which particle motion stops; also known as 0 Kelvin.

acid: a chemical that donates protons (H+ ions) when dissolved in water.

active transport: uses energy to move molecules across a cell membrane against a concentration gradient.

adipose tissue: found beneath the skin and around organs, providing cushioning, insulation, and fat storage.

aerobic: steps in the cellular respiration process that require oxygen.

alpha decay: occurs when the nucleus of an atom emits a package of two protons and two neutrons, called an alpha particle, which is equivalent to the nucleus of a helium atom.

alveoli: thin-walled air sacs, which are the site of gas exchange.

anabolism: the process whereby cells build molecules and store energy.

anaerobic: steps in the cellular respiration process that do not require oxygen.

anaphase: when the centromere divides, chromatids are separated from each other and become a chromosome; the two identical chromosomes move along the spindle fibers to opposite ends of the cell.

anions: negative ions.

Archimedes' principle: when an object is placed in a fluid, the object will have a buoyant force equal to the weight of the displaced fluid.

arteries: larger vessels that carry blood away from the heart.

atmospheric pressure: pressure that results from the total weight of the atmosphere exerting force on the Earth; can be measured with a barometer.

atom: the simplest unit of an element that retains the element's characteristics.

atomic mass: calculated by adding up the masses of the protons and neutrons.

atomic number: the number of protons found in the nucleus of an atom of that element.

atomic weight: the average mass number.

B cells: class of lymphocyte cells that merge from the bone marrow mature and produce antibodies, which enter the bloodstream.

base: a chemical that accepts protons (H+ ions) when dissolved in water.

beta decay: occurs when the nucleus emits a beta particle that degrades into an electron as it passes out of the atom.

binomial nomenclature: methodology proposed by Linnaeus that used two Latin-based categories—genus and species—to name each organism, which can be further classified based on family, order, class, phylum and kingdom.

biogeography: the study of how photosynthetic organisms and animals are distributed in a particular location, plus the history of their distribution in the past.

biome: an ecosystem that is generally defined by its climate characteristics.

biosphere: part of the Earth that includes all living things.

blood tissue: flows through the blood vessels and heart and is essential for carrying oxygen to cells, fighting infection, and carrying nutrients and wastes to and from cells.

boiling point: temperature at which a substance changes from liquid to gas.

capillaries: tiny vessels that surround all tissues of the body and exchange carbon dioxide for oxygen.

cardiac muscle: tissue forming the walls of the heart with strength and electrical properties that are vital to the heart's ability to pump blood.

carrying capacity: within a given area, there is a maximum level the population may reach at which it will continue to thrive.

cartilage tissue: reduces friction between bones and supports and connects them.

catalyst: a substance that changes the speed of a reaction without being affected itself.

cations: positive ions.

cell: the smallest and most basic unit of most living things.

cell cycle: a particular sequence of events ending in cell division, which produces two daughter cells.

cell division: the process of cell reproduction that centers on the replication and separation of strands of DNA.

cell membrane: structure that encloses the cell and separates it from the environment; also known as the plasma membrane.

cell walls: made up of cellulose and lignin, they enclose the cell membrane providing strength and protection for the cell.

cellular metabolism: a general term that includes all types of energy transformation processes, including photosynthesis, respiration, growth, movement, etc.

central nervous system (CNS): two main components, the brain and the spinal cord, which control all other organs and systems of the body.

cerebellum: part of the brain that controls balance, equilibrium, and muscle coordination.

cerebrum: part of the brain that controls sensory and motor responses, memory, speech, and most factors of intelligence.

chemistry: the study of matter.

chlorophyll: pigment molecules that give the chloroplasts their green color.

chloroplasts: the site of photosynthesis within plant cells.

chromatids: the two identical strands of duplicated chromatin in a cell that is getting ready to divide.

chromatin: the combination of DNA with histones.

circuit: the path that an electric current follows.

circulatory system: the conduit for delivering nutrients and gases to all cells and for removing waste products from them.

CO2 fixation: the second phase of photosynthesis in which six CO_2 molecules are linked with hydrogen (produced in photolysis), forming glucose (a six-carbon sugar); also known as the dark reaction.

community: populations that interact with each other in a particular ecosystem.

competition: results when two or more species living within the same area and that overlap niches both require a resource that is in limited supply.

compound: formed when two or more different atoms bond together chemically to form a unique substance.

condensation: change of a gaseous substance to liquid form.

conditioning: involves learning to apply an old response to a new stimulus.

conduction: movement of energy by transfer from particle to particle; can only occur when objects are touching.

connective tissue: holds tissues and organs together, stabilizing the body structure.

cortex: a ring inside the epidermis that is made up of large parenchyma cells.

covalent bond: bond formed between atoms when atoms share electrons.

cytoplasm: region between the nucleus and cell membrane.

cytoskeleton: provides structural support to a cell.

density: the measure of how much matter exists in a given volume.

differential reproduction: individuals within a population that are most adapted to the environment and are also the most likely individuals to reproduce successfully; tends to strengthen the frequency of expression of desirable traits across the population.

diffraction: the bending of a light wave around an obstacle.

diffusion: the process whereby molecules and ions flow through the cell membrane from an area of higher concentration to an area of lower concentration; mixing of particles in a gas or liquid.

digestion: breakdown of ingested particles into molecules that can be absorbed by the body.

digestive system: serves as a processing plant for ingested food.

diploid: the parent cell that has a normal set of paired chromosomes.

dispersion: process in which a species may move in or out of a particular area over the course of time.

displacement: measures the change in position of an object, using the starting point and ending point and noting the direction.

division: distributes the remaining set of chromosomes in a mitosis-like process.

domains: classification category even more general than kingdoms.

dominance: occurs when older, more established individuals compete for status within the community.

dominant allele: an allele that masks the effect of its partner allele.

ecology: the study of how organisms interact with other organisms and how they influence or are influenced by their physical environment.

ecosystem: a group of populations found within a given locality, plus the inanimate environment around those populations.

egg cell: produced by the female gametophyte; also referred to as a female gamete.

electrical current: a flow of electrons through a conductor.

electron cloud: three-dimensional space where electrons travel freely; also known as an electron shell or orbital.

electron transport: the second step of aerobic respiration that captures the energy created by the release of electrons from the Krebs cycle.

element: a substance that cannot be broken down into any other substances.

embryo: what a zygote eventually grows into.

emigration: permanent one-way movement out of the original range.

endocytosis: the process whereby large molecules are taken up into a pocket of membrane; the pocket pinches off, delivering the molecules, still inside a membrane sack, into the cytoplasm.

endoderm: the precursor of the gut lining and various accessory structures.

endothermic: reactions that require energy.

energy cycle: supports life throughout the environment.

enzymes: protein molecules that act as catalysts for organic reactions.

epithelial tissue: forms the barrier between the environment and the interior of the body.

eukaryotic cells: cells that contain membrane-bound intracellular organelles, including a nucleus.

evaporation: escape of individual particles of a substance into gaseous form.

evolution: the gradual change of characteristics within a population, producing a change in species over time.

excretory system: responsible for collecting waste materials and transporting them to organs that expel them from the body.

exocytosis: the export of substances from the cell.

exothermic: reactions that release energy.

extinction: when the entire population of a particular species is eliminated.

F1 generation: Mendel's first generation of offspring.

fermentation: another name for anaerobic respiration, which breaks down the two pyruvic acid molecules (three carbons each) into end products (such as ethyl alcohol, C_2H_6O or lactic acid $C_3H_6O_3$), plus carbon dioxide (CO_2).

fertilization: occurs when two haploid cells join to form a diploid cell.

flower: the primary reproductive organ for a plant.

food chain: energy generally flows through the entire ecosystem in one direction from producers to consumers and on to decomposers.

force: the push or pull exerted on an object.

forebrain: located most anterior, it contains the olfactory lobes and cerebrum as well as the thalamus, hypothalamus, and pituitary gland.

freezing point: temperature at which a substance changes from liquid to solid.

frequency: the number of wavelengths that pass a point in a second.

friction: the rubbing force that acts against motion between two touching surfaces.

gametes: the four haploid cells (egg and sperm) that are found in reproductive organs as a result of meiosis.

gametophytes: generated by the reproductive organs of the sporophyte through the process of meiosis.

gamma radiation: consists of gamma rays, which are high-frequency, high-energy, electromagnetic radiation that are usually given off in combination with alpha and beta decay.

gas exchange system: responsible for the intake and processing of gases required by an organism and for expelling gases produced as waste products; also known as the respiratory system.

gene: length of DNA that encodes a particular protein.

gene migration: the introduction of new genes from an immigrant, which results in a change of the gene pool.

gene pool: the entire collection of genes within a given population.

genetic drift: over time, a gene pool (particularly in a small population) may experience a change in frequency of particular genes simply due to chance fluctuations.

genetic engineering: the intentional alteration of genetic material of a living organism.

genomes: sum total of genetic information.

genotype: the combination of alleles that make a particular trait.

glycolysis: the breaking down of the six-carbon sugar (glucose) into smaller carbon-containing molecules yielding ATP.

Golgi apparatus: instrumental in the storing, packaging, and shipping of proteins; also known as Golgi bodies or the Golgi complex.

habitat: the physical place where a species lives.

habituation: a learned behavior in which the organism produces less and less response as a stimulus is repeated, without a subsequent negative or positive action.

half-life: the time it takes for 50 percent of an isotope to decay.

Hardy-Weinberg law of equilibrium: in situations where random mating is occurring within a population (which is in equilibrium with its environment), gene frequencies and genotype ratios will remain constant from generation to generation.

heat: energy that flows from an object that is warm to an object that is cooler.

hemoglobin: component of blood responsible for carrying oxygen.

heterozygous: when the two alleles for a given gene are different in an individual.

hindbrain: consists of the cerebellum and medulla oblongata.

histones: short length of DNA wrapped around a core of small proteins.

homeostasis: a state of dynamic equilibrium, which balances forces tending toward change with forces acceptable for life functions.

homologous: structures that exist in two different species because they share a common ancestry.

homozygous: when both alleles for a given gene are the same in an individual.

hormones: chemicals produced in the endocrine glands of an organism, which travel through the circulatory system and are taken up by specific targeted organs or tissues, where they modify metabolic activities.

hydrogen bond: occurs when a hydrogen atom is involved with a polar intermolecular attraction to a more electronegative atom.

hypothalamus: involved in hunger, thirst, blood pressure, body temperature, hostility, pain, pleasure, etc.

immigration: permanent one-way movement into a new range.

immune system: functions to defend the body from infection by bacteria and viruses.

imprinting: a learned behavior that develops in a critical or sensitive period of the animal's lifespan.

inheritance: the process by which characteristics pass from one generation to another.

innate behaviors: the actions in animals we call instincts; highly stereotyped.

insulators: poor conductors of electrical current.

invertebrates: those species having no internal backbone structure.

ionic bond: bond of attraction between positive and negative ions.

ions: charged atoms.

isotopes: atoms with the same number of protons but different numbers of neutrons.

kidneys: filter metabolic wastes from the blood and excrete them as urine.

Krebs cycle: the first step in aerobic respiration that occurs in the matrix of a cell's mitochondria and breaks down pyruvic acid molecules (three carbons each) into CO_2 molecules, H+ (protons), and 2 ATP molecules; also liberates electrons.

law of dominance: one gene is usually dominant over the other.

law of inertia: a particle at rest will stay at rest and a particle in motion will stay in motion until acted upon by an outside force.

law of segregation: traits are expressed from a pair of genes in the individual (on homologous chromosomes).

laws of thermodynamics: explain the interaction between heat and work (energy) in the universe.

learned behaviors: may have some basis in genetics, but they also require learning.

lymphatic system: the principal infection-fighting component of the immune system.

lymphocytes: begin in bone marrow as stem cells and are collected and distributed via the lymph nodes.

magnetism: the ability of a substance to produce a magnetic field.

mass: the amount of matter that is contained by the object.

mechanics: the study of things in motion.

medulla oblongata: part of the brain that controls involuntary responses such as breathing and heartbeat.

meiosis: the process of producing four daughter cells, each with single unduplicated chromosomes.

melting point: temperature at which a substance changes from solid to liquid form.

Mendel, Gregor: studied the relationships between traits expressed in parents and offspring and the hereditary factors that caused expression of traits.

metaphase: occurs when the spindle fibers pull the chromosomes into alignment along the equatorial plane of the cell, creating the metaphase plate.

midbrain: between the forebrain and hindbrain and contains the optic lobes.

migration: temporary movement out of one range into another and back.

mitochondria: centers of cellular respiration.

mitosis: the process by which a cell distributes its duplicated chromosomes so that each daughter cell has a full set of chromosomes.

molar mass: the mass in grams of one mole of atoms.

molecule: two or more atoms held together by shared electrons (covalent bonds).

momentum: the product of mass and velocity.

mRNA: RNA strand that migrates from the nucleus to the cytoplasm; also known as messenger RNA.

musculoskeletal system: provides the body with structure, stability, and the ability to move.

mutation: a change of the DNA sequence of a gene, resulting in a change of the trait.

natural selection: a feature of population genetics that is the driving force behind evolution.

nerve tissue: carries electrical and chemical impulses to and from organs and limbs to the brain.

nervous system: a communication network that connects the entire body of an organism and provides control over bodily functions and actions.

neurons: carry impulses via electrochemical responses.

Newton's laws of motion: three laws that form the basis of most of our understanding of things in motion.

nuclear membrane: the boundary between the nucleus and the cytoplasm.

nucleolus: a rounded area within the nucleus of the cell where ribosomal RNA is synthesized.

nucleus: an organelle surrounded by two lipid bilayer membranes that is located near the center of the cell and contains chromosomes, nuclear pores, nucleoplasm, and nucleoli.

olfactory lobes: responsible for the sense of smell.

oogenesis: formation of egg cells.

optic lobes: visual center connected to the eyes by the optic nerves.

organelles: cell components that perform particular functions.

organic compounds: the building blocks of all living things.

organism: an individual of a particular species.

osmosis: a special process of diffusion that occurs when the water concentration inside the cell differs from the concentration outside the cell; the water on the side of the membrane with the highest water concentration will move through the membrane until the concentration is equalized on both sides.

ovary: the hollow, bulb-shaped structure in the lower interior of the pistil.

ovules: small round cases within the ovary that contain one or more egg cells.

P1 generation: Mendel's term for the first generation of true-breeding plants; also known as the parent.

parasympathetic nervous system: carries impulses back from organs.

Pascal's principle: the pressure exerted on any point of a confined fluid is transmitted unchanged throughout the fluid.

passive transport: substances freely pass across the membrane without the cell expending any energy.

periodic table: listing of elements by atomic number.

peripheral nervous system (PNS): a network of nerves throughout the body.

pH: potential of hydrogen scale, which is a measurement of H+ ions in solution.

phenotype: the trait expressed.

photosynthesis: a crucial set of reactions that convert the light energy of the Sun into chemical energy usable by living things.

pituitary gland: releases various hormones.

placenta: in mammals, the structure that forms when the outer cells of the embryo and the inner cells of the uterus combine.

polar molecules: molecules that have regions of partial charge.

polygenic traits: traits produced from interaction of multiple sets of genes.

population: the total number of a single species of organism found in a given ecosystem.

population density: the number of individuals of a particular species living in a particular area.

positron decay: occurs when the nucleus emits a particle that degrades into a positron as it passes out of the atom.

pressure: a measure of the amount of force applied per unit of area.

prokaryotic cells: cells with no nucleus or any other membrane-bound organelles.

prophase: the first stage of mitosis; chromatin condenses into chromosomes within the nucleus, the centrioles move to opposite ends of the cell, and spindle fibers begin to extend from the centromeres of each chromosome toward the center of the cell.

proteins: present in every living cell, large unbranched chains of amino acids; may also be called polypeptides.

Punnett square: notation that allows us to easily predict the results of a genetic cross.

quantum mechanics: predicts the probabilities of an electron being in a certain area at a certain time.

radiation: the transfer of energy via waves.

reactants: reacting molecules.

recessive allele: the allele that does not produce its trait when present with a dominant allele.

reflection: the bouncing of a wave of light off an object.

reflexes: an automatic movement of a body part in response to a stimulus.

refraction: the change of direction of a wave as it passes from one medium to another.

regulation: enzyme control that may occur when the product of the reaction is also an inhibitor to the reaction.

ribosomes: the site of protein synthesis within cells.

S phase: second phase of interphase where cell begins to prepare for cell division by replicating the DNA and proteins necessary to form a new set of chromosomes.

sex-limited trait: genes that are located on a gender chromosome.

sex-linked traits: more males develop the trait because males have only one copy of the X chromosome; females have a second X gene, which may carry a gene coding for a functional protein for the trait in question, which may counteract a recessive trait.

skeletal muscle: attaches bones of the skeleton to each other and surrounding tissues, which enables voluntary movement.

skin: an accessory excretory organ that secretes wastes with water from sweat glands.

smooth muscle: makes up the walls of internal organs and functions in involuntary movement (breathing, etc.).

social behavior: behavior patterns that take into account other individuals.

society: an organization of individuals in a population in which tasks are divided in order for the group to work together.

somatic motor nerves: carry impulses to skeletal muscle from the CNS.

somatic sensory nerves: carry impulses from body surface to the CNS.

special theory of relativity: Einstein's theory that states that the speed of light is a constant and the laws of physics are the same in all inertial (non-accelerating) reference frames.

specific heat: the measure of a substance's ability to retain energy.

speed: the rate of change of an object's distance traveled.

sperm: produced by the male gametophyte; also known as a male gamete.

stroma: the body of the chloroplast.

structural genes: code proteins that form organs and structural characteristics.

substrate: particular substance of an enzyme that fits within the active site.

succession: when one community completely replaces another over time in a given area.

symbiosis: when two species interact with each other within the same range.

sympathetic nervous system: carries impulses that stimulate organs.

synapse: point at which homologous chromosomes pair up during meiosis.

T cells: mature cells in the thymus gland that patrol the blood for antigens but are also equipped to destroy antigens themselves.

taxonomy: study that organizes living things into groups based on morphology or, more recently, genetics.

telophase: occurs as nuclear membranes form around the chromosomes and disperse through the new nucleoplasm; spindle fibers also disappear.

temperature: the measure of the average kinetic energy of a substance.

territory: an area of land that lies within the home range that the individual will defend as his own.

thymus: a mass of lymph tissue that is active only through the teen years, fighting infection and producing T cells.

transcription: the formation of an RNA molecule, which corresponds to a gene.

transduction: the transfer of genetic material (portions of a bacterial chromosome) from one bacterial cell to another.

transformation: a process in which bacteria absorb and incorporate pieces of DNA from their environment (usually from dead bacterial cells).

translation: phase of photosynthesis that requires a second type of RNA.

transpiration: a process in which some water that has traveled up through the plant to the leaves is evaporated.

tRNA: a chain of about 80 nucleotides that provide the link between the "language" of nucleotides (codon and anticodon) and the "language" of amino acids; also known as transfer RNA.

valence shell: the outermost occupied energy level of an element.

Van der Waals forces: momentary forces of attraction that exist between molecules and are much weaker than the forces of chemical bonding.

vascular: plants that have tissue organized in such a way as to conduct food and water throughout their structure; also known as tracheophytes.

vector: mathematical quantities that recognize both the size and direction of the dimension being considered.

veins: vessels that carry blood toward the heart.

velocity: the rate of change of displacement; includes both speed and direction.

vertebrates: species that have internal backbones.

vessels: arteries, veins, and capillaries.

villi: protrusions out into the lumen of the intestine that provide a large surface area for absorption of nutrients.

visceral sensory nerves: carry impulses from body organs to the CNS.

voltage: the electromotive force that pushes electrons through the circuit.

Watson-Crick model: named after scientists James Watson and Francis Crick, who discovered and modeled the structure of DNA.

wavelength: the distance from one crest (or top) of a wave to the next crest on the same side.

weight: the force of gravity acting upon that object.

work: the movement of a mass over a distance.

zygote: cell that results when a sperm cell fertilizes an egg cell.

Index

A

Abiotic, 94
Abiotic factors, 100
Absolute zero, 123
Absolute zero, law of, 124
Acceleration, 132
Accessory organs, 72
Active site, 58
Active transport, 43, 4536
Adaptive radiation, 21, 27
Adipose tissue, 71
Aerobic respiration, 47
Aganatha, 31
Alimentary canal, 72
Alleles, 52, 84
Allopatric speciation, 26–27
Alpha decay, 112
Alpha particle, 112
Alveoli, 74
Amensalism, 102
Ammonification, 97
Amperage, 137
Amphibia, 31
Anabolism, 46
Anaerobic reaction, 48
Analogous, 21
Anaphase, 54, 57
Angiosperms, 63, 64–67
Animal cells, 37–41
Animals, 69–83
  circulatory system, 78–79
  digestive system, 72–73
  evolution of, 20–22
  excretory system, 79–80
  homeostatic mechanisms, 81
  hormonal control in homeostasis and reproduction, 81–82
  immune system, 80–81
  musculoskeletal system, 74–75
  nervous system, 75–78
  reproduction, 82–83
  respiratory or gas exchange system, 73–74
  tissue of, 70–71
  traits of, 69–70
Anion, 118
Annelida, 30
Annuals, 63
Anther, 65
Antibodies, 80
Antigens, 80
Anus, 73
Aquatic biomes, 104
Archaea, 29
Archimedes' principle, 127
Arteries, 78, 79
Arterioles, 79
Arthropoda, 31
Assortment, law of independent, 86–87
Asteroids, 149
Asthenosphere, 162
Astronomy, 145–153
  Big Bang, 146–147
  galaxies, 145
  moon, 151–153
  solar system, 147–151
  stars, 146–147
  Sun, 147
Atmosphere, 93, 157–160, 161
Atmospheric pressure, 160
Atomic chemistry. *See* Chemistry
Atomic mass, 111
Atomic number, 111
Atomic weight, 111
Atoms, 109–112
ATP (adenosine triphosphate), 46
Attached ribosomes, 39
Auroras, 160
Australopithecus afarensis, 22
Autotrophs, 95
Aves, 31
Axon, 75

B

Barometer, 160
B cells, 80
Beta decay, 113
Beta particle, 113
Biennials, 63
Big Bang, 146–147
Bile, 80
Binomial nomenclature, 28
Bioenergetics, 45–46
Biogeochemical cycles, 94
Biomass, 95
Biomes, 104
Biosphere, 93
Biosynthesis, 57–59
Biotic factors, 94, 100
Birth rate, 99
Black hole, 145
Blood, 78–79
Blood tissue, 71
Boiling point, 126
Bone tissue, 71
Botany. See Plants
Brain, 75, 76
Bronchi, 74
Bronchioles, 74
Buoyancy, 127

C

Cambrian explosion, 21
Cambrian period, 20–21
Capillaries, 78, 79
Carbon cycle, 97–98
Carbon dioxide, 98
Cardiac muscles, 71, 75
Carnivores, 95
Carrying capacity, 15, 101
Cartilage tissue, 71
Catabolism, 46
Cation, 118
Cell body, 75
Cell cycle, 53–57
  G_1 phase, 53
  G_2 phase, 53
  interphase, 53
  meiosis, 55–57
  mitosis, 53–55
  S phase, 53
Cell division, 51–57
Cell membrane, 36, 37, 42
  properties of, 43–45
Cells
  aerobic respiration, 47
  animal, 37–41
  ATP, 46
  cellular respiration, 47–48
  cycle of, 53–57
  defined, 35
  division of, 51–57
  DNA replication, 49–51
  energy transformations, 45–46
  evolution of first, 16–18
  meiosis, 55–57
  mitosis, 53–55
  organelles, 36–42
  photosynthesis, 46–47
  photosynthetic, 36
  plant, 41–42
  properties of cell membranes, 43–45
  shape and size of, 37
  types of, 35–36
Cellular metabolism, 46
Cellular respiration, 47–48
Cell walls, 42
Central Nervous System, 76
Central vacuole, 42
Centrioles, 39
Centromere, 52
Cephalochordata, 31
Cerebellum, 77–78
Cerebrum, 77
Ceres, 148
Chemical bonds, 118–119
Chemical reactions, 119–120
Chemistry, 109–114
  attractions between molecules, 118–119
  chemical bonds, 118–119
  chemical reactions, 119–120
  common elements, 117
  half-life, 114
  nuclear reactions and equations, 112–113

periodic table, 110–111
properties of water, 119
structure and physical properties of substances, 118–119
structure of atom, 109–112
Chlorophyll, 42, 47
Chloroplasts, 42
Chondrichthyes, 31
Chordata, 31
Chromatids, 52
Chromatin, 51
Chromosomes, 84
   structure of, 51–52
Circuits, 137–138
Circulatory system, 78–79
Class (classification), 28–29
Classical mechanics, 129–132
Classification of living organisms, 28–32
Climax community, 104
Closed circuit, 137
Closed circulatory system, 78
Closed community, 103
Cnidaria, 30
CO_2 fixation, 47
Coacervates, 17
Codon, 50
Cohesion, 68
Cohesion-tension process, 68
Combination, chemical reaction, 119
Community, 93
Community structure, 103–104
Competitive exclusion, 102
Compounds, 118
Condensation, 126
Conduction, 123
Conductors, 137
Connective tissue, 70
Conservation of matter and energy, law of, 124
Convection, 123
Convergent evolution, 21–22
Core, Earth's, 162
Covalent bonds, 118
Crescent moon, 151, 152
Cristae, 40
Cro-Magnon man, 22
Crossing over, 56, 89

Crust, Earth's, 161
Cuticle, 67
Cytokinesis, 55
Cytoplasm, 36
Cytoplasmic organelles, 36
Cytoskeleton, 39

D

Darwin, Charles, 15
Death rate, 99
Decomposition, 119
Denitrification, 97
Density
   population, 101
   of substances, 126–127
Density-dependent factors, 101
Density-independent factors, 101–102
Differential reproduction, 24
Diffraction, 141
Diffusion, 43, 126
Digestion, 72
Digestive system, 72–73
Dihybrid cross, 86
Diploid, 55
Diploid phases, 69
Dispersion, 102
Displacement, 130
Division, 55
DNA (deoxyribonucleic acid)
   mutation, 51
   replication of, 49–51
   structural and regulatory genes, 51
   structure of, 48–49
Domains, 29
Dominance, law of, 86
Dominant trait, 84
Dwarf planets, 148

E

Earth, 148, 157–164
   atmosphere of, 157–160
   core, 162
   crust, 161
   facts about, 157
   geologic column, 163–164

hydrosphere, 163
magnetosphere, 163
mantle, 162
Mohorovicic and Gutenberg discontinuities, 162
tile of, and seasons, 150–151
Echinodermata, 31
Ecology, 93–99
  biomes, 104
  carbon cycle, 97–98
  ecological cycles, 94
  energy cycle, 94–96
  nitrogen cycle, 96–97
  phosphorous cycle, 98–99
  population growth and regulation, 99–104
  terms for, 93–94
  water cycle, 96
Ecosystem, 93
Ecotones, 103
Egestion, 72
Egg cell
  animal, 83
  plants, 69
Einstein, Albert, 133–134
Electrical current, 137
Electricity, 137–138
  series and parallel circuits, 137–138
  static, 137
Electromagnetic radiation, 112
Electron cloud, 111
Electron microscope, 36
Electrons, 109, 137
  characteristics of, 111
Electron shell, 112
Elements
  defined, 109
  most common elements, 117
  periodic table, 110–111
Embryo
  animal, 82, 83
  plants, 66, 69
Emigration, 102
Endocytic vesicles, 37
Endocytosis, 45
Endoplasmic reticulum, 39
Endosperm, 66

Endosymbiont hypothesis, 40
Endosymbiont theory, 18–19
Endothermic reaction, 120
Energy, 137–141
  electricity, 137–138
  of electromagnetic radiation, 112
  magnetism, 138
  waves, 139–141
Energy cycle, 94–96
Energy level, 112
Energy transformations, 45–46
Entropy, law of, 120, 124
Environment, 93
Enzymes
  defined, 57
  naming, 58
  reactions of, 58–59
Enzyme-substrate complex, 58
Epidermis, leaves, 67
Epiglottis, 74
Epithelial tissue, 70
Eris, 148
Esophagus, 72
Eubacteria, 29
Eukaryota, 29
Eukaryotic cells, 35
Evaporation, 126
Event horizon, 145
Evolution, 15–32
  of animals, 20–22
  convergent, 21–22
  differential reproduction, 24
  of first cells, 16–18
  genetic drift, 24
  Hardy-Weinberg Law of Equilibrium, 25–26
  of humans, 22
  mechanisms of, 23–26
  mutation, 24
  natural selection, 15–16
  of plants, 19–20
  speciation, 26–28
Excretory system, 79–80
Exocytosis, 40, 45
Exosphere, 159–160
Exothermic reaction, 120
Extinction, 22, 103

F

F_1 generation, 83
F_2 generation, 83
Facilitated diffusion, 43, 44
Family (classification), 28–29
Feces, 73
Feedback control, 81
Feedback response, 81
Female gamete, 69
Fertilization, 69, 82
Filament, 65
Flower, 64–65
Food chain, 94–95
Food chain pyramid, 95
Food web, 96
Force, 132
Forebrain, 77
Fossil fuels, 98
Fox, Sidney, 17
Free ribosomes, 39
Freezing point, 126
Frequency, 139
Friction, 131
Fruits, 66
Full moon, 151, 152
Fusion, heat of, 126

G

G_1 phase, 53
G_2 phase, 53
Galaxies, 145
Gall bladder, 73
Gametes, 55
Gametogenesis, 82
Gametophytes, 69
Gamma radiation, 113
Gamma rays, 113
Gases, 124–126
Gas exchange system, 73–74
Gas Giants, 149
Gastrointestinal tract, 72
Gene migration, 25
Gene pool, 23–24
Genes, 48–51
  defined, 49
  DNA replication, 49–51
  mutation, 51
  regulatory, 51
  structural, 51
Genetic drift, 24
Genetics, 83–90
  genotype, 84
  incomplete dominance, 87–88
  law of dominance, 86
  law of independent assortment, 86–87
  law of segregation, 86
  linkage, 88–89
  Mendel's discoveries, 83–84
  modern genetics, 84–86
  multiple alleles, 88
  phenotype, 84
  polygenic inheritance, 90
  Punnett square, 84–85
Genomes, 48
Genotype, 24, 84
Genus, 28–29
Geologic column, 163–164
Geologic time scale, 23, 164
Geosphere, 160
Glottis, 74
Glycolysis, 47–48
Gnathostomata, 31
Golgi apparatus, 40, 41
Grana, 42
Gravitational constant, 128
Gravity, 127–129
Gray matter, 78
Greenhouse gas, 96
Guard cells, 67
Gutenberg discontinuity, 162
Gymnosperms, 63

H

Habitat, 94
Halley's comet, 150
Half-life, 114
Haploid, 55
Haploid phases, 69
Hardy, G. H., 25
Hardy-Weinberg Law of Equilibrium, 25–26

Heart, 78
Heat, 123–124
  defined, 123
  of fusion, 126
  specific, 124
  of vaporization, 126
Hemoglobin, 79
Hereditary. *See* Genetics
Heterozygous, 85
Hindbrain, 77
Histones, 51
Homeostasis, 81
  in ecosystems, 101
Homeostatic mechanisms, 81
Homo erectus, 22
Homologous, 21
Homologous chromosomes, 84
Homologs, 52
Homo sapiens, 22, 28
Homozygous, 85
Hormone-receptor complex, 82
Hormones, 81–82
Human evolution, 22
Hydrogen bonds, 119
Hydrologic cycle, 96
Hydrosphere, 93, 163
Hypothalamus, 77

I

Immigration, 102
Immune system, 80–81
Incomplete dominance, 87–88
Ingestion, 72
Inheritance, 83
Inner core, 162
Insulators, 137
Interference, 139
Interphase, 53
Intestines, 73
Invertebrates, 70
Ionic bond, 118
Ionosphere, 160
Ions, 118
Isotopes, 111

J

Jupiter, 148, 149

K

Kidneys, 79
Kinetic energy, 123
Kinetochore, 54
Kingdom, 28–29
Kingdom Animalia, 29–31
Kingdom Fungi, 29–30
Kingdom Plantae, 29–30
Kingdom Protista, 29–30

L

Large intestines, 73
Larynx, 74
Leaves, 66–67
Light, 141
Light spectrum, 141
Limiting factors of populations, 100–103
Linkage, 88–89
Linnaeus, Carolus, 28
Liquids, 124–126
Lithosphere, 93, 162
Liver, 73, 80
Locus, 84
Longitudinal waves, 139
Lunar eclipse, 153
Lungs, 74, 80
Lymph, 80–81
Lymphatic system, 80–81
Lymph nodes, 80
Lymphocytes, 80
Lysosomes, 40

M

Magnetism, 138
Magnetosphere, 163
Male gamete, 69
Mammalia, 32
Mantle, 162
Mars, 148
Mass
  atomic, 111
  defined, 127
Mass number, 111
Matter, states of, 124–126

Mechanics
  classical, 129–132
  defined, 129
Medulla oblongata, 78
Meiosis, 55–57
Melting point, 126
Mendel, Gregor, 15, 83
Mercury, 148
Mesopause, 159
Mesosphere, 159
Metaphase, 54, 56–57
Metric units, 132
Microfilaments, 39
Microtubules, 39
Microvilli, 38
Midbrain, 77
Migration, 102
Milky Way, 145
Miller, Stanley, 17
Minimal viable population, 103
Mitochondria, 40, 41
Mitosis, 53–55
Mobile receptor mechanism, 81
Moho, 162
Mohorovicic discontinuity, 162
Molar mass, 111
Molecules
  attractions between molecules, 118–119
  defined, 118
  polar, 118–119
Mollusca, 30
Momentum, 132
Monohybrid cross, 86
Moon, 151–153
Mortality, 99
Mouth, 72
mRNA, 50
Multicellular, 35–36
Muscles, 74–75
Muscle tissue, 71
Musculoskeletal system, 74–75
Mutation, 24
  DNA replication, 50–51
Mutualism, 102
Myelin sheath, 78

N

Nasal passages, 74
Natality, 99

Natural selection
  Darwinian concept of, 15
  modern concept of, 15–16
Nematoda, 30
Neptune, 148, 149
Nerve tissue, 71
Nervous system, 75–78
Neurons, 37, 75
Neutrons, 109
New moon, 151, 152
Newton, Isaac, 128
Newton's laws of motion, 129–130
Niche, 94
Nitrates, 97
Nitrification, 97
Nitrogen cycle, 96–97
Nitrogen fixing, 97
Nonvascular plants, 63
Nose, 74
Nuclear membrane, 41
Nuclear pores, 41
Nuclear reactions and equations, 112–113
Nucleolus, 40, 42
Nucleosomes, 51
Nucleus, 36, 40, 42

O

Olfactory lobes, 77
Omnivores, 95
Oocyte, 83
Oogenesis, 83
Oparin, A. I., 16
Oparin Hypothesis, 16–17
Open circuit, 137
Open circulatory system, 78
Open community, 103
Optic lobes, 77
Orbital, 112
Order (classification), 28–29
Ordovician period, 21
Organelles, 35
Organism, 93
Organisms, 35
Osmosis, 44
Osteichthyes, 31
Outer core, 162
Ovary, 65, 83
Ovules, 65

P

P_1 generation, 83
Paleozoic Era, 20
Palisade layer, 67
Pancreas, 73
Parallel circuits, 138
Parasitism, 102–103
Parasympathetic nervous system, 76
Parenchyma cells, 68
Pascal's principle, 127
Passive transport, 43–44
Pedicel, 65
Perennials, 63
Periodic table, 110–111
Peripheral Nervous System, 76
Petals, 65
Pharynx, 74
Phenotype, 24, 84
Phloem, 66
Phosphorus cycle, 98–99
Photolysis, 47
Photosynthesis, 46–47
Photosynthetic cells, 36
Phylogenetic tree, 29
Phylum, 28–29
Physics, 123–134
  classical mechanics, 129–132
  density, 126–127
  electricity, 137–138
  gravity, 127–129
  heat, 123–124
  laws of thermodynamics, 124
  magnetism, 138
  Newton's laws of motion, 129–130
  solids, liquids, gases and plasma, 124–126
  specific heat, 124
  states of matter, 124–126
  theory of relativity, 132–134
  waves, 139–141
Pioneer communities, 104
Pistil, 65
Pituitary gland, 77
Planets, 147–151
Plant cells, 41–42
Plants, 63–69
  anatomy of, 63–67
  angiosperms, 63, 64–67
  asexual plant reproduction, 69
  evolution of, 19–20
  food translocation and storage, 68–69
  growth and development, 69
  gymnosperms, 63
  nonvascular, 63
  reproduction, 69
  vascular, 63
  water and mineral absorption and transport, 68–69
Plasma, 124–125
Plasma membrane, 37
Plasmodesmata, 68
Platyhelminthes, 30
Ploidy, 55
Pluto, 148
Polar body, 83
Polar molecules, 118–119
Pollen grains, 65–66
Pollen tube, 66
Polygenic inheritance, 90
Ponnamperuma, Cyril, 18
Population growth and regulation, 99–104
  community structure, 103–104
  limiting factors, 100–103
  models of, 99
  population defined, 93
Porifera, 30
Positron decay, 113
Post-transcriptional processing, 50
Predator/prey relationship, 102
Pressure, 126–127
Primary oocytes, 83
Primary spermatocytes, 82
Primordial soup, 17
Products, chemical reactions, 119
Prokaryotes, 18
Prokaryotic cells, 35, 36
Prophase, 54, 56
Protons, 109
Punctuated equilibrium, 27
Punnett square, 84–85

Q

Quantum mechanics, 111–112

R

Radiation, 123
  adaptive, 21
Range, of species, 101–102
Reactants, 119
Reactions, chemical, 119
Receptor, 82
Recessive trait, 84
Rectum, 73
Reduction, 55
Reflection, 139, 141
Refraction, 141
Regulatory genes, 51
Relativity, theory of, 132–134
Replacement reaction, 120
Reproduction
  animals, 82–83
  plants, 69
Reptilia, 31
Resistance, 137
Respiratory system, 73–74
Ribosomes, 39, 42
  translation, 50
Root system, 68
Rough endoplasmic reticulum, 39, 41

S

Salivary glands, 72
Saturn, 148, 149
Secondary oocyte, 83
Secondary spermatocytes, 82
Secretory vesicles, 40
Seeds, 66
Segregation, law of, 86
Sensory organs, 75
Sepals, 64–65
Series circuits, 138
Sex-influenced trait, 89
Sex-limited trait, 89
Sex-linked trait, 89
Sieve plates, 69
Silurian period, 21
SI units, 132
Skeletal muscles, 71, 75
Skin, 80
Small intestines, 73
Smooth endoplasmic reticulum, 39, 42
Smooth muscles, 71, 75
Solar system, 147–151
Solids, 124–126
Somatic motor nerves, 76
Somatic sensory nerves, 76
Sound waves, 140–141
Speciation, 26–28
Species, 28–29
Specific heat, 124
Speed, 130–131
Sperm, 69
Spermatids, 82
Spermatogenesis, 82
Sperm cells, 82
Sperm nuclei, 66
S phase, 53
Spinal cord, 76
Spindle fibers, 54
Spleen, 80
Spongy layer, 67
Sporophyte, 69
Stars, 146–147
States of matter, 124–126
Stem, 66
Steroid, 81
Stigma, 65
Stimulus, 81
Stomach, 72
Stomata, 67
Stratosphere, 158–159
Stroma, 42
Structural genes, 51
Style, 65
Substrate, 58
Succession, 104
Sun, 147, 149
Symbiosis, 102
Sympathetic nervous system, 76
Sympatric speciation, 27
Synapse, 56
Synapses, 75
Systema Naturae (Linnaeus), 28
Système Internationale (SI) units, 132

T

Taxonomy, 28
T cells, 80–81
Telophase, 55
Temperature, 123
Terminal velocity, 129
Terrestrial biomes, 104
Thalamus, 77
Thermodynamics, laws of, 120, 124
Thermopause, 159
Thermosphere, 159
Thymus, 80
Tonoplast, 42
Tonsils, 80
Trachea, 74
Transcription, 50
Translation, 50
Transpiration, 68
Transverse wave, 139
Trilobites, 21
tRNA, 50
Trophic levels, 95
Troposphere, 158
Tube nucleus, 66

U

Unicellular, 35
Universe, 145–153
   Big Bang, 146–147
   galaxies, 145
   moon, 151–153
   solar system, 147–151
   stars, 147
   Sun, 147
Uranus, 148, 149
Urine, 79–80
Urochordata, 31

V

Valence shell, 112
Vaporization, heat of, 126
Vascular bundle, 67
Vascular plants, 63
Vascular tissue, 66
Vectors, 130
Vegetative propagation, 69
Veins, 78, 79
Velocity, 131
   terminal, 129
Venules, 79
Venus, 148
Vertebrata, 31
Vertebrates, 70
Vessels, 78
Villi, 73
Viruses, 35
Visceral sensory nerves, 76
Voltage, 137

W

Wallace, Alfred Russell, 15
Water
   absorption and transport in plants, 68–69
   properties of, 119
Water cycle, 96
Wavelength, 139
Waves, 139–141
   light, 141
   longitudinal, 139
   properties of, 140
   sound, 140–141
   transverse, 139
Weight, 127–128
Weinberg, Wilhelm, 25
White matter, 78
Work, 130

X

Xylem, 66
Xylem cells, 68

Z

Zeroth Law of Thermodynamics, 124
Zoology. *See* Animals
Zygote, 66, 69, 82